Rock Mechanics

Felsmechanik

Mécanique des Roches

Supplementum 9

Tectonic Stresses
in the Alpine-Mediterranean Region

Proceedings of the Symposium
Held in Vienna, Austria, September 13-14, 1979

Under the Auspices
of the Working Group 3 of the Interunion Commission
of Geodynamics and the European Geophysical Society

Edited by
Adrian E. Scheidegger

1980   Springer-Verlag   Wien   New York

Prof. Dipl.-Phys. Dr. phil. Adrian E. Scheidegger
Institute of Geophysics
Technical University
Wien, Austria

With 122 Figures

ISSN 0080-3375
ISBN-13:978-3-211-81578-6   e-ISBN-13:978-3-7091-8588-9
DOI: 10.1007/978-3-7091-8588-9

Preface

The idea of the present symposium was conceived a few years ago by the Working Group 3 (WG3) of the Interunion Commission of Geodynamics (ICG). Now that its Program has been completed, it was decided to present the results of that part of the activities of the WG3 indicated in the symposium title.

The writer was asked by the WG3 to coordinate the contributions. The European Geophysical Society (EGS) kindly offered to host the symposium during its meeting in Vienna in September, 1979, and Springer-Verlag in Vienna agreed to publish a "Proceedings" volume of the Symposium. These efforts are gratefully acknowledged.

The whole subject matter of the symposium was divided into 6 "themes". The first concerns in situ stress determinations, the second deals with stresses as inferred from fault-plane solutions of earthquakes, the third geomorphic and geological effects of stresses, the fourth petrofabrics and stresses, the fifth recent displacements and the sixth geomechanical models.

Upon the recommendation of the WG3, contributors were invited to present papers falling under the themes mentioned. The symposium, however, was also open to unsolicited contributors in the general call for papers for the EGS-meeting. It is hoped that the papers will present a fairly complete picture of the results obtained in the study of tectonic stresses in the Alpine-Mediterranean region during the ten years of the duration of the International Geodynamics Program.

A. E. Scheidegger

Contents

VI   Contents

Theme 4
Petrofabrics and Stresses

Theme 5
Recent Displacements

Theme 6
Geomechanical Models

Rock Mechanics, Suppl. 9, 1—3 (1980)

Rock Mechanics
Felsmechanik
Mécanique des Roches
© by Springer-Verlag 1980

Theme 1

In Situ Stress Measurements

Stress Fields, Fracture Systems and the Mechanism for Movements in the Gneiss-Granite Area of the Mont Blanc Massif

By

Nils Hast

With 1 Figure

Abstract

Results of in situ stress measurements performed in bore-holes from the road tunnel between Chamonix and Courmayeur are given.

The middle part of the 11.5 km long road tunnel through the Mont Blanc Massif between Chamonix and Courmayeur passes through a 7 km long wide area of gneiss-granite (Fig. 1). Along a plane (A – B) sloping at an angle of 65^0 to the horizontal, this area is bounded parallel to the Chamonix Valley in France by a 3 km-wide zone of crystalline schist, and at the Italian end of the tunnel by a 1.5 km-wide zone of limestone.

In a comprehensive project carried out under my supervision by the Swedish Research Laboratory, "Rock Stress Measurements AB", Stockholm, the stress conditions in the young, still active, Alpine mountain range were investigated with a view to exploring the cause of movements taking place in the rock of the Massif and thus obtaining an insight into the way in which mountains are formed.

The location of the long tunnel at depths below ground level — at places as much as 2500 m — provided a unique opportunity for such a study by means of in-situ stress measurements in boreholes from the tunnel face. Over a period of nearly one year (1970—1971) measurements of the absolute values of stress in the bedrock around the tunnel were formed at a large number of points in boreholes drilled from turning bays. The bore-holes were drilled to such a depth that the presence of the tunnel could not exert more than a negligible effect on the determination of the prevailing stresses.

In different parts of the gneiss-granite area at the level of the tunnel the *horizontal compressive stress* are uniform in magnitude, about 65 MPa, while the *shearing stress* horizontally in the vertical plane about 7 MPa.

0080-3375/80/Suppl. 9/0001/$ 01.00

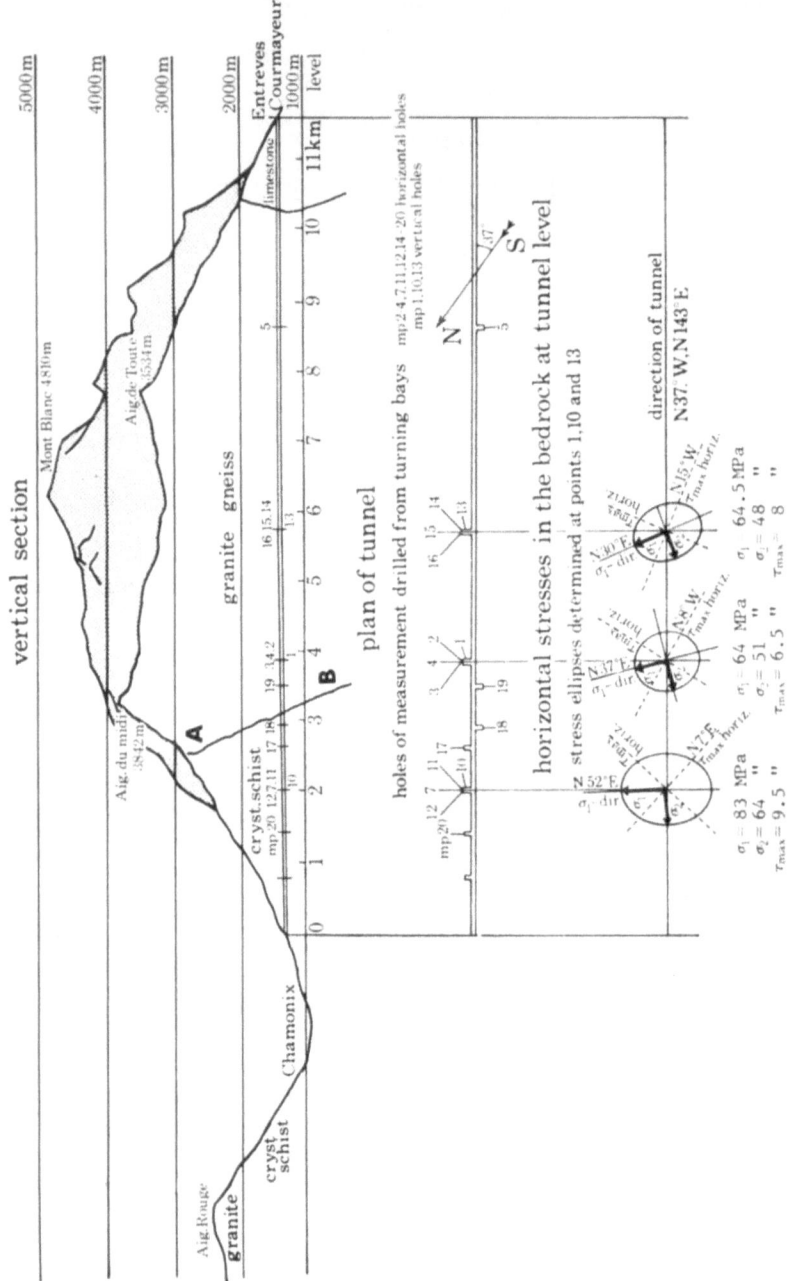

Fig. 1. Cross-section through Mt. Blanc Massif, showing stresses

In the triangular crystalline schist area facing the Chamonix Valley the values in the tunnel are higher, with peaks of about 80 MPa for the compressive and 9.5 MPa for the shear stress. The directions of τ_{max} vary somewhat between N 15⁰ W and N 6⁰ E ($\pm 90^0$ in both cases); the mean thus does not deviate greatly from the N−S and E−W directions; these directions of fracturing are manifested very clearly in peaks such as Adolfo Ray.

The *vertical compressive stress* in both the granite-gneiss and the crystalline schist, on the other hand, fluctuates considerably from one part of the Massif to another. In part of the gneiss-granite area, 660 m from the plane A−B in Fig. 1, the vertical compressive stress is 80 MPa, but falls off to 60 MPa at the middle of the area and to 40 MPa at a point 2600 m from the Italian end of the tunnel. Considerable changes were also found in the maximum shear stress in the vertical plane. At the corresponding points where the measurements were made, the values of the vertical shear recorded were 13, 6.5 and 4 MPa. In the schist area and 2 km from the French end of the tunnel there is a maximum vertical compressive stress of about 80 MPa; τ_{max} here is 14 MPa. From here both these magnitudes decreased towards the end of the tunnel and toward the zone of contact with the gneiss-granite zone.

The predominant directions of fractures in the gneiss-granite zone belong to a system of vertical orthogonal fracture planes which, with their N−S and E−W directions unchanged, penetrate the Massif from the tunnel level up to the mountain peaks. The fractures in question seem, however, to have been produced by shear forces not at the level of the tunnel, but rather deep in the crust where the shear stresses are of a much greater magnitude.

An analysis of the measurement results provides clear evidence that the anomalies in the vertical stress field in the area compared to the normal field in a bedrock have their origin in movements deep in the crust. The gneiss-granite block is pressed upwards by vertical forces, but this movement is resisted by friction in the plane A−B between the block and the crystalline schist. This situation results in vertical shearing forces which vary in magnitude from one part of this area to another and in the excess of the overburden weight in the rock.

Thus, it is demonstrated by the in-situ stress measurements that the gneiss-granite block in the Mont Blanc Massif is still undergoing upward displacement.

Address of author: Prof. Nils Hast, Engelbreksgatan 16, S-11432 Stockholm, Sweden.

Rock Mechanics, Suppl. 9, 5—15 (1980)

Rock Mechanics
Felsmechanik
Mécanique des Roches
© by Springer-Verlag 1980

Tectonic Stresses in the Northern Foreland of the Alpine System Measurements and Interpretation

By

G. Greiner and J. Lohr

With 7 Figures

Abstract

The regional stress field in the northern foreland of the Alps has been investigated by *in situ* stress determinations. More than 1000 strain relief measurements were made with resistance strain gages in boreholes carried out in mines, tunnels and deep quarries. The stresses calculated and data from other authors were used to get a detailed idea of the stress field in the region under consideration. The measurements confirm a continuous trend of compressive stresses from the Alps to the northern foreland. The largest stresses were observed in the Central Alps, the lowest in the Rhinegraben and, surprisingly, in the central part of the Rhenish Massif. The horizontal stresses exceed the vertical ones at most places. This observed behaviour of the stresses has been checked by other data from seismic surveys and measurements in deep oil and gas wells of the Molasse basin north of the Alps.

1. Introduction

During the last decennium tectonic investigations in the foreland of the Alpine fold belt have been accelerated by several national and international programs. One of the main problems under consideration was the neotectonic behaviour of the earth's crust in Central Europe north of the Alps.

This area is cut by an active rift belt. The main segments of it are the Rhinegraben, a seismic active zone through the Rhenish Massif and the Lower Rhine Embayment with the adjacent Zuider Zee depression (Netherlands). All together, these elements form a subplate boundary with an extension of about 800 km (Illies, 1977; Illies and Greiner, 1979) that may be traced not only by surface geologic features but also by an accumulation of seismic events along the path of this main tectonic line (Ahorner, 1975; Bonjer and Fuchs, 1979).

Some of the earthquakes like these of Basel 1356 and Rastatt 1933 were destructive (Illies and Greiner, 1978; Schmitt-Zittel, 1933) and demonstrate clearly that strong tectonic forces are still at work in this region.

To understand more of these remarkable crustal deformations, a detailed program for mapping tectonic stresses has been started in Germany; some of the results are presented here.

0080-3375/80/Suppl. 9/0005/$ 02.20

2. Measuring Technique

At the test sites the strain-relief technique was used for determining the absolute stresses in the interior of the rock masses. This technique has primarily been developed in South Africa and is known as the "doorstopper strain gage device" (Leemann, 1969).

The principle of this overcoring technique is demonstrated in Fig. 1: A hole is drilled into the rock and at its flattened end an oriented rosette of strain gages is glued. At that time the stresses in the rock mass are still present. After initial readings the installation tool is taken away and the

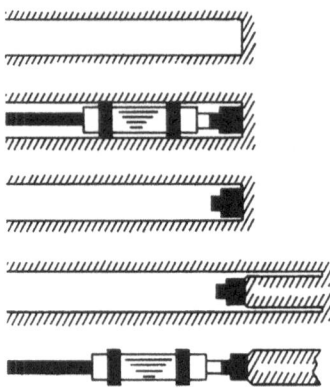

Fig. 1. Schematic illustration of the overcoring method for the determination of *in situ* stress by strain relief

strain gage is overcored. The result of this operation is a radial extension of the core with the strain gage on it. The strain will be proportional to the amount of stress release in the rock when overcoring.

The radial strains are determined by difference readings at the gages outside the borehole. Using Hooke's law it is possible to calculate the virgin stress tensor in the rock mass around the borehole from the determined strain tensor.

The test sites were situated in mines, tunnels and deep quarries which, after careful examination, were found to have the optimal conditions with regard to position, topography, geometric feature, rock type and geologic structure. These ideal circumstances were possible for a lot of the measurements have been done for scientific purposes only. Nevertheless, the application of all the stress determination techniques to tectonic stress measurements requires caution as a number of secondary effects, such as thermal, residual, gravitational and shape-induced stresses can also be present in the rock mass. A detailed discussion of these problems was given by Greiner (1978).

At each test site more than 60 strain readings were taken in order to obtain sufficient data for statistical evaluations. This means that the following investigations are based upon more than 1300 separate strain values taken in a variety of rocks in different tectonic environments by the Karlsruhe working group (Illies and Greiner, 1978).

When measuring in quarries, several boreholes were drilled at different places into the quarry floor. Continuous measurements with increasing depths up to more than 10 m have shown that near-surface measurements are worthless, since boundary stresses as well as thermally induced stress-variations influence such measurements severely. This is consistent with the

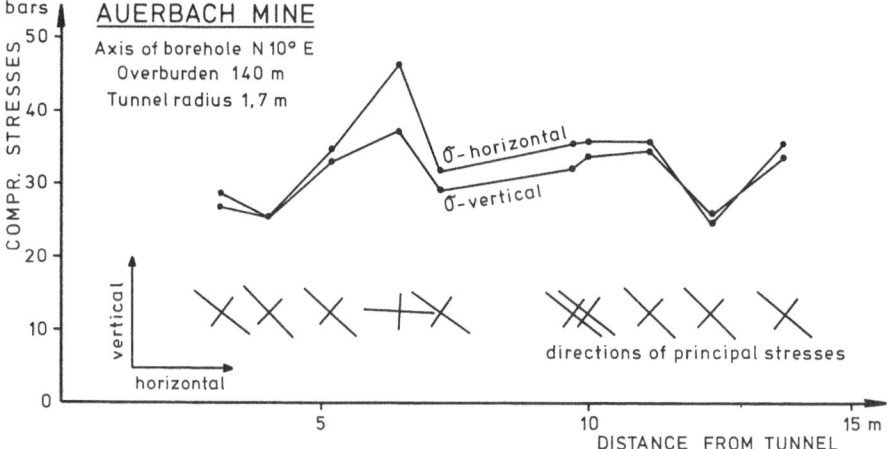

Fig. 2. Variation of stresses in a bore-hole with increasing depth (10 bar = 1 MPa)

results of Hooker and Duvall (1971) and Lee and Nichols (1972), who have shown that diurnal thermally induced stresses reach down to 1—2 meters whilst annual temperature variations reach at least to a depth of 5—6 m. This gives a clear indication of the required depths for obtaining significant data.

Fig. 2 gives an example of the data quality when measuring in a tunnel without thermal stress effects. The measurements were performed continuously in several boreholes up to depths of 25 m. This means that the last measurements were taken at least at a distance of about 7—8 times the radius of the tunnel away from the tunnel wall.

The measurements presented were taken in dioritic rocks at a depth of 140 m. The scatter of the directions of the principal stresses is very small with one exception in the vicinity of a fracture zone at 6.5 m. The standard deviation of the stress magnitudes is about 12% when neglecting the extremely high values near the fracture zone.

3. Results and Interpretation

The results of our measurements are shown in Fig. 3. In addition to this set of data, results from other authors have been used: (References given by Greiner and Illies, 1977).

The area investigated is characterized by a relatively uniform orientation of the directions of maximum compression: a general NNW – SSE-trend was found in the Alps. This trend is confirmed by new measurements of Kohlbeck (1979) and Martinetti and Ribacchi (1979) from measurements in undisturbed areas.

The region north of the Alps is characterized by the same orientation of the maximum horizontal stress but it is obvious that there is a slight

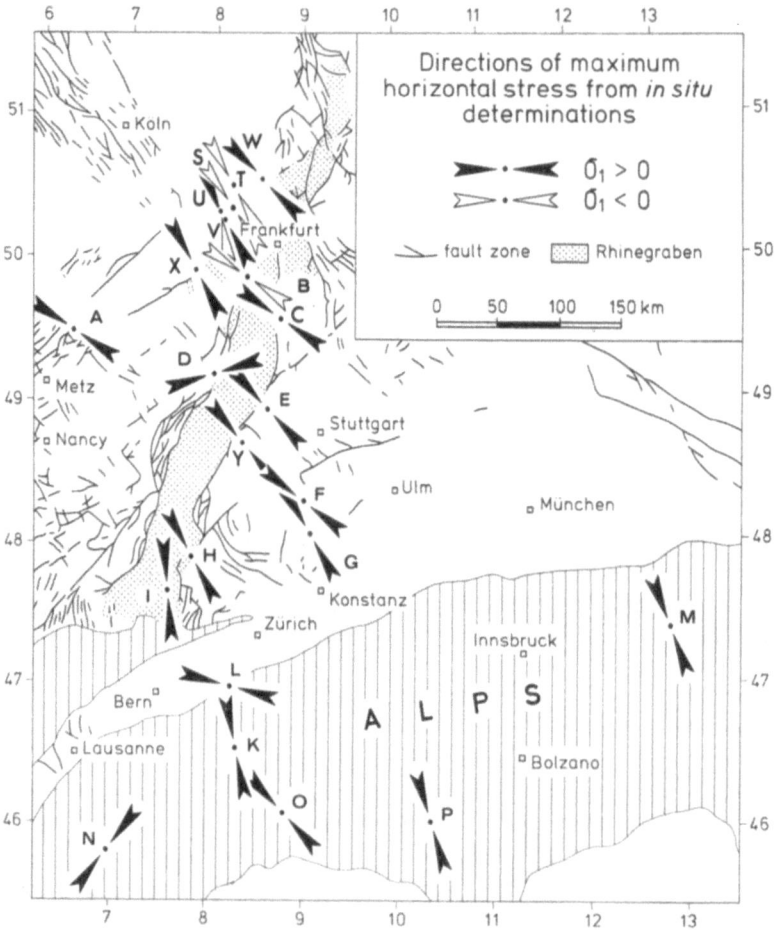

Fig. 3. Directions of maximum horizontal compressive stress according to *in situ* measurements. The dominant direction is northwest to north-northwest within the Alps and in the foreland

rotation from values of about 160—170⁰ in the southern part of the Rhinegraben to 120—130⁰ in the northern part of it. In the Rhenish Massif the values become larger again.

Within this context it is important to note that the tectonic structure of the Rhinegraben has no significant effect upon the directions, for some local disturbances may well be explained by near-surface or residual stresses.

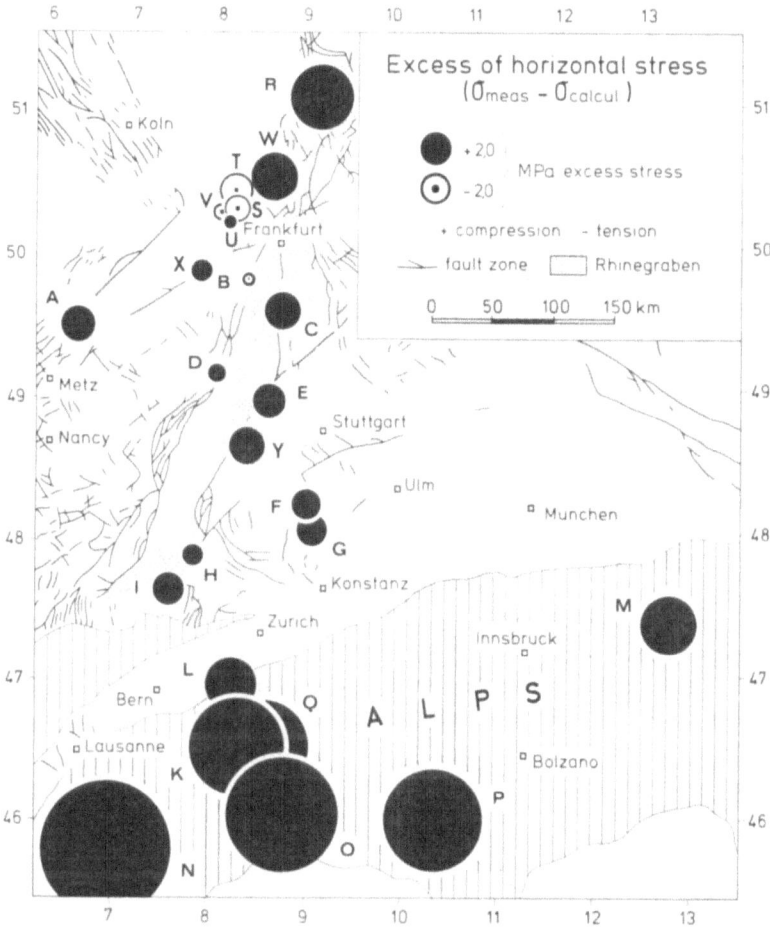

Fig. 4. Magnitude of the excess horizontal stress. The stresses are relatively low in the foreland, increase along the northern margin of the Alps, and culminate in the central part of the mountain range (10 bar = 1 MPa)

Not only do the results presented here confirm this statement, but measurements on the other side of the Vosges Mountains in France also gave corresponding values for the directions and values of the stresses (Paquin et al., 1978).

Contrary to the approximately uniform distribution of the directions, the magnitudes of the stresses show larger variations (Fig. 4): In order to compare the values of the tectonically most important horizontal stresses from measurements at different depths, the excess of horizontal stress was

calculated as the difference of the measured horizontal stress and the theoretical horizontal stress (Poisson's stress) that should be present, considering
the lithostatic stress caused by the weight of the overburden and the resulting horizontal compression due to the lateral expansion of the rocks
(Greiner and Illies, 1977). These values were plotted on a tectonic map
of Central Europe to show the interrelations of stresses and tectonic features.

In the investigated area, a systematic distribution of the stresses is present.
According to neotectonic processes, the stress field may be divided into
three major units: Alpine System, Rhinegraben, Rhenish Massif.

The Alpine System shows very high stresses, in some places more than
35 MPa. From the northern rim of the fold belt up to the Central Alps a
systematic increase of the horizontal compressive stresses is observed. The
culmination should be in the region where the highest recent uplift of about
1.7 mm/year is recorded (Gubler, 1976).

In the adjacent northern foreland blocks, the magnitude drops drasti-
cally to an average value of about 2 MPa. Within the foreland blocks the
Rhinegraben area shows a minimum of the compressive stresses. Away from
the flanks of the graben, the magnitudes increase again. Stresses with a
negative magnitude (extensions) were observed near the northern termination of the graben to the Rhenish Massif.

This extensional character of the stress field is much clearer in the
Rhenish Massif itself (Schmidt, 1979). Especially along the axis of a seismically active zone connecting the Rhinegraben and the fault trough of the
Lower Rhine Embayment (Ahorner, 1975), the stresses are very low, at
some places even negative. At some distance from this zone the stresses
increase significantly; they are higher than those in the foreland of the Alps
but lower than those determined in the mountain range itself.

The same stress behaviour is reflected in data from other surveys made
in this area, especially in the region of the Alpine System. For example,
there are records of very high horizontal stresses in the Bavarian subalpine
Molasse. Here the walls of deep galleries in a lignite mine had been strongly
deformed by thrusting processes (Heissbauer, 1975). Investigations made
in the same mining district during local earthquake swarms confirmed the
tendency of man-triggered active overthrust faulting (Illies and Greiner,
1978). Moreover, the hydrocarbons accumulated in different strata of the
Molasse trough are found to be under abnormally high overpressures that
increase significantly towards the rim of the fold belt (Lemcke, 1973).
These high stresses are thought to be the result of lateral compression from
the still active Alpine System (Lemcke, 1978). In the same way excess pres-
sures of sealed-in ground water in deep-seated aquifers are observed as
evidenced by gas exploration wells (Müller et al., 1975).

The most important data for the evaluation of the stresses in the
northern foreland may be derived from the seismic velocities in the tertiary
sediments of the Molasse trough (Lohr, 1967, 1978): Fig. 5 shows the velocity-depth functions for the Molasse trough and the Rhinegraben. Obviously, the velocities increase with depth much faster in the western part
of the sedimentary basin than in the eastern part and in the Rhinegraben,

although the lithology within the areas under consideration can approximately be taken as uniform. In addition to the compaction due to gravitational stresses (Gardner et al., 1974), a strong further effect must be present to produce these differences in the velocity-depth functions.

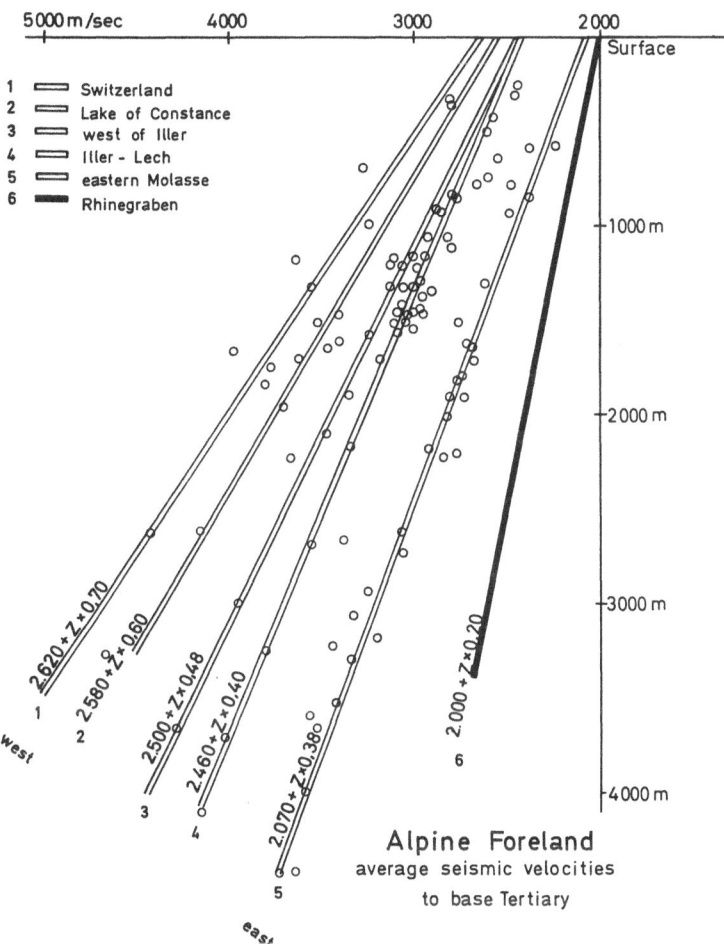

Fig. 5. Molasse through and Rhinegraben. Average seismic velocities down to the basis of the Tertiary

An impression of the behaviour of the seismic velocities as determined by velocity measurements in deep oil and gas wells is given in Fig. 6. Considering the velocity-depth function of the Rhinegraben as a scale it is possible to map the excess velocities in the Molasse trough north of the Alps. In this context it is necessary to confirm that the increase of velocity with depth seems to be very small in the Rhinegraben so that the excess velocities become relatively high. Nevertheless, the excess velocities are a good means in the investigation of neotectonic interactions.

The picture shows clearly an increasing excess stress when approaching the Alpine fold belt. The lines of equal excess velocities do not follow the morphology of the sedimentary trough which indicates additional stress effects as well as the spacing of the isolines. In Switzerland where the Molasse trough is smallest, the highest excess velocities are observed.

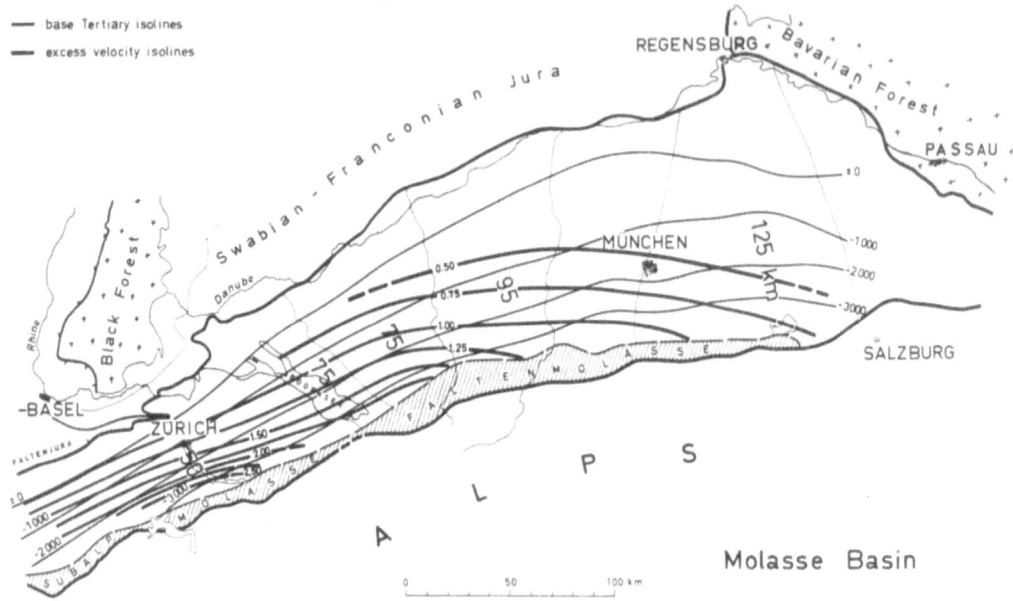

Fig. 6. Lines of equal depth of the basis of the Tertiary and isolines of excess velocities. Highest velocities near the northern rim of the Subalpine Folded Molasse

The dependence of basin-width to excess velocities is documented in Fig. 7 where there is a clear relationship between basin width, e. g. distance from the Alps and the seismic velocities. Similar behaviour of the velocities was observed in the cretaceous sediments in the eastern foreland of the Rocky Mountains in Southern Alberta, Canada (Davis, 1975). There, like in the Molasse trough, high lateral compression is thought to cause these anomalies by increasing compaction affecting especially the elastic moduli (Gardner et al., 1974). This is consistent with *in situ* stress determinations that show the highest values exactly at the places where the excess velocities become largest, e. g. where the Molasse trough is narrowest and the Variscian massif of the Black Forest acts as an abutment.

Lohr (1978) gives an estimation of the additional lateral stresses of about 46 to 76 MPa that are required to create the velocity-anomalies. This agrees with the values measured *in situ* within the Alpine System.

The last question concerning these investigations refers to the origin of the stresses. The maximum horizontal stresses are recorded in the Central Alps, from here they are transferred to the foreland. Surely these stresses are not residual stresses from former orogenetic processes (Greiner and Illies, 1977). The main parts of the stresses must be active tectonic stresses of plate tectonic origin. Corresponding to these tectonic stresses, a straining

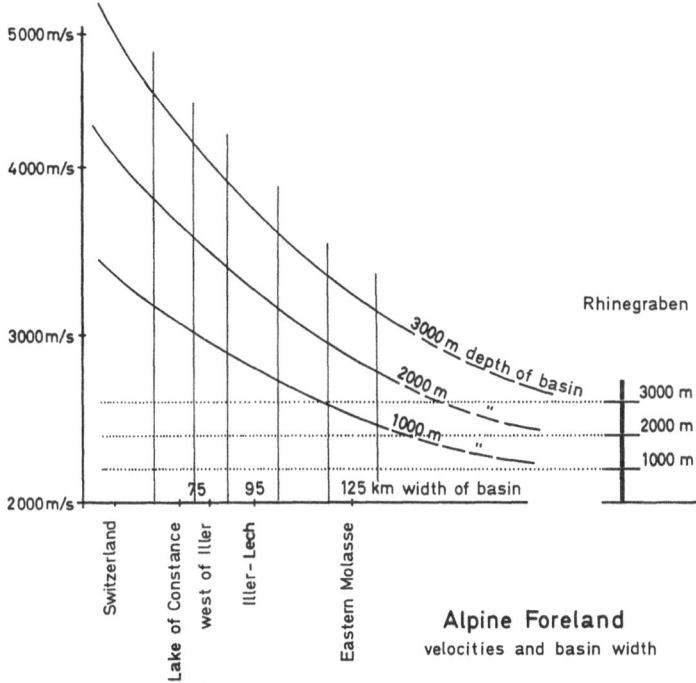

Fig. 7. Relationship between velocities and width of basin. Highest velocities in the smallest part of the Molasse trough

of the foreland is observed that is expressed by systems of active conjugate shears and young fracture deformations (Pavoni and Mayer-Rosa, 1978; Scheidegger, 1977).

The topographic-gravitational effect of the mountain itself as well as phase transformations of minerals within the mountain body due to post-folding uplift and denudation (Neugebauer et al., 1976) are not strong enough to build up the observed stress field. It is most likely a combination of all these effects with a clear dominance of active intraplate tectonic stresses.

References

Ahorner, L.: Present-day Stress Field and Seismotectonic Block Movements Along Major Fault Zones in Central Europe. Tectonophysics 29, 233—249 (1975).

Bonjer, K.-P., Fuchs, K.: Real-time Monitoring of Seismic Activity and Earthquake Mechanisms in the Rhinegraben Area as a Basis for Prediction. Proc. ESA-Council of Europe, Seminar on Earthquake Prediction (Strasbourg, March 5—7, 1979). SP 149, 57—62 (1979).

Davis, T. L.: Influence of Lateral Compression on Seismic Velocity in Southern Alberta. Paper given at 45th Intern. S. E. G. Meeting, Denver, 21 p. (1975).

Gardner, G. H. F., Gardner, L. W., Gregory, A. R.: Formation Velocity and Density — A Diagnostic Basics of Stratigraphic Traps. Geophysics 39, 770—780 (1974).

Greiner, G.: Spannungen in der Erdkruste — Bestimmung und Interpretation am Beispiel von In-situ-Messungen im süddeutschen Raum. Universität Karlsruhe (TH) Diss. 1978, 192 p. (1978).

Greiner, G., Illies, J. H.: Central Europe: Active or Residual Tectonic Stresses. Pageoph 115, 11—26 (1977).

Gubler, E.: Beitrag des Landesnivellements zur Bestimmung vertikaler Krustenbewegungen in der Gotthard-Region. Schweiz. Mineral. Petrogr. Mitt. 569, 675—678 (1976).

Heissbauer, H.: Die Gebirgsmechanik beim Abbau in großer Teufe des Kohlenbergwerkes Peißenberg und ihre Auswirkung auf die Bergtechnik. Geologica Bavarica 73, 37—53 (1975).

Hooker, V. E., Duvall, W. I.: In Situ Rock Temperature. Stress Investigations in Quarries. Bureau of Mines Rept. of Invest. 7589, 1—12 (1972).

Illies, J. H.: Ancient and Recent Rifting in the Rhinegraben. Geol. Mijnbouw 56, 329—350 (1977).

Illies, J. H., Greiner, G.: Rhinegraben and the Alpine System. Geol. Soc. Am. Bull. 89, 770—782 (1978).

Illies, J. H., Greiner, G.: Holocene Movements in the Rhinegraben Rift System. Tectonophysics 52, 349—359 (1979).

Kohlbeck, K., Roch, K. H., Scheidegger, A. E.: In Situ Stress Measurements in Austria. This Volume.

Lee, F. T., Nichols (Jr.), T. C.: Some Effects of Geologic Structure, Engineering Operations and Thermal Changes on Rockmass Behaviour. 24th I. G. C. Montreal, Sect. 13, 261—272 (1972).

Leeman, E. R.: The "Doorstopper" and Triaxial Rock Stress Measuring Instruments Developed by the CSIR. J. S. Afr. Inst. Min. Metall. 69, 305—339 (1969).

Lemcke, K.: Zur nachpermischen Geschichte des nördlichen Alpenvorlandes. Geologica Bavarica 69, 5—48 (1973).

Lemcke, K.: Summary of Post-Permian History in the Northern Alpine Foreland. Alps, Apennines, Hellenides. Inter-Union Commission on Geodynamics, Scientific Rpt. 38, 61—64 (1978).

Lohr, J.: Die seismischen Geschwindigkeiten der jüngeren Molasse im ostschweizerischen und deutschen Alpenvorland. Geophys. Prosp. 17, 111—125 (1969).

Lohr, J.: Alpine Stress Documented by Anomalous Seismic Velocities in the Molasse Trough. Alps, Apennines, Hellenides. Inter-Union Comission on Geodynamics, Scientific Rept. 38, 69—71 (1978).

Martinetti, S., Ribacchi, R.: In Situ Stress Determinations in Italy. This Volume (1979).

Müller, M.: Bohrung Miesbach 1: Ergebnisse der ersten im Rahmen des Erdgastiefenaufschlußprogrammes der BRD mit öffentlichen Mitteln geförderten Erdgastiefbohrung. 3. DGMK-Fachgruppentagung Hannover, 63—76 (1975).

Neugebauer, H. J., Brötz, R., Rybach, L.: On the Dynamics of the Swiss Alps Along the Geotraverse Basel–Chiasso. Schweiz. Mineral. Petrogr. Mitt. *56*, 703—706 (1976).

Paquin, Ch., Froidevaux, C., Souriau, M.: Mesures directes des contraintes tectoniques en France septentrionale. Bull. Soc. géol. France, *1978* (7), t. 20, No. *5*, 727—731 (1978).

Pavoni, N., Mayer-Rosa, D.: Seismotektonische Karte der Schweiz 1:750000. Eclogae geol. Helv. *71*, 293—295 (1978).

Scheidegger, A. E.: Kluftmessungen im Gelände und ihre Bedeutung für die Bestimmung des tektonischen Spannungsfeldes in der Schweiz. Geographica Helv. *32* (3), 121—134 (1977).

Schmidt, T. J.: In Situ Stress Profile Through the Alps and Foreland. Allgem. Vermessungs-Nachrichten *86* (10), 367—376 (1979).

Schmidt-Zittel, H.: Vorläufige Mitteilung über das Rastatter Erdbeben vom 8. Februar 1933. Bad. Geol. Abh. *5*, 140—158 (1933).

Address of authors: G. Greiner, J. Lohr, Gewerkschaften Brigitta und Elwerath Betriebsführungsgesellschaft m. b. H., D-3000 Hannover, Federal Republic of Germany.

Rock Mechanics, Suppl. 9, 17—18 (1980)

Rock Mechanics
Felsmechanik
Mécanique des Roches
© by Springer-Verlag 1980

Tectonic Stresses in France

By

C. Paquin and C. Froidevaux

With 1 Figure

Abstract

A flat jack method has been used to determine the principal stress axes in the Jurassic limestone of the basin of Paris and in neighbouring regions. The results have been compared with a few overcoring measurements in boreholes of a few meters.

The principal compression axis is NNW so that the stress field appears to be very homogeneous from the Rhine graben to the Atlantic coast. Comparisons with neotectonic studies and earthquake mechanisms confirm this conclusion for the whole of France with the exception of the Alpine belt.

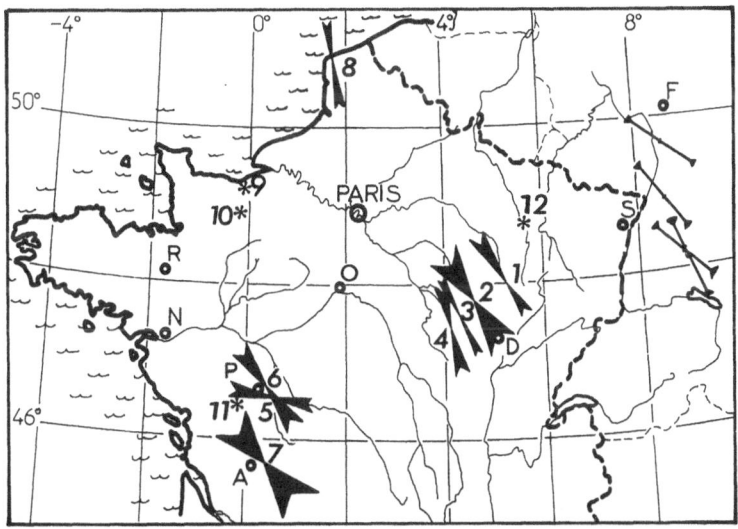

Fig. 1. Map showing in-situ stresses in France

In situ measurements of horizontal stresses have been carried out by the flat jack method. This requires flat rock surfaces of good mechanical quality. Limestone quarries in the Jurassic of the Parisian and Acquitaine Basins

0080-3375/80/Suppl. 9/0017/$ 01.00

were found to be adequate. The map in Fig. 1 shows our results (Paquin et al., 1978): arrows indicate the azimuth of the major horizontal stress, the angular extension of each arrow indicating the 95% confidence interval. The arrows in the Rhine graben are from Greiner (1975). The results show a very homogeneous stress field from the Rhine graben to the Atlantic coast with a principal axis of compression $\sigma_{H\max}$ in the NNW direction. This can be interpreted in terms of an African "push" on the European plate. The stress $\sigma_{H\max}$ is typically 2 MPa (20 bars) at the surface, whereas $\sigma_{H\min}$ is much smaller and often extensive.

References

Greiner, G.: In Situ Stress Measurements in South West Germany. Tectonophysics 29, 265—274 (1975).

Paquin, C., Froidevaux, C., Souriau, M.: Mesures directes des contraintes tectoniques en France septentrionale. Bull. Soc. Géol. France 20, 135—139 (1978).

Address of authors: C. Paquin, C. Froidevaux, Laboratoire de Géophysique et Géodynamique interne, Université Paris Sud, F-91405 Orsay, France.

Rock Mechanics, Suppl. 9, 19—20 (1980)

Rock Mechanics
Felsmechanik
Mécanique des Roches
© by Springer-Verlag 1980

In Situ Stress Measurements in Greece

By

C. Froidevaux, C. Paquin, C. Angelidis, and A. Tzitziras

With 1 Figure

Abstract

This paper presents the results of in situ stress measurements in Greece.

Greece is one of the most seismically active regions of Europe. Its tectonic setting within the Alpine belt has been interpreted as that of an area of continuous deformation. In the West one has a zone of continent-continent collision, in the South West and South a zone of convergence with an island arc. The push of the Arabian peninsula on the Asian plate results in

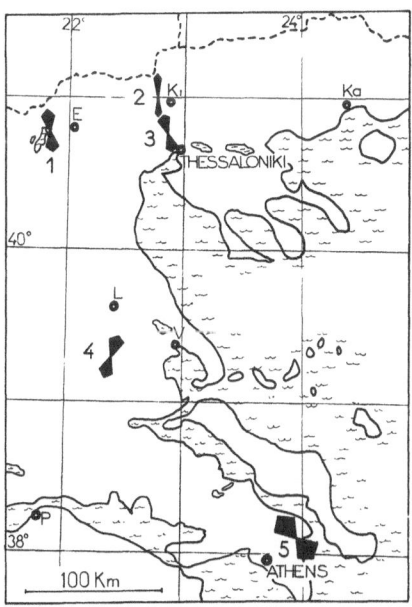

Fig. 1. Azimuth of the least horizontal stress $\sigma_{H\min}$. The angular domain of each arrow indicates the scatter in directions measured at a given site. All $\sigma_{H\min}$ are extensional. A single measurement performed close to site 2 yielded identical results. About 50 km east of site 1 a site was investigated and yielded a compressive $\sigma_{H\min}$

2*

0080-3375/80/Suppl. 9/0019/$ 01.00

a westward motion of the Turkish peninsula inducing large scale intraplate deformations within the whole Agean domain. The 1978 Thessaloniki earthquake has been analysed in terms of a stress pattern with major extension in the North-South direction (Mercier, 1979). According to earthquake mechanisms and neotectonic studies, a similar stress field should be found in most of continental Greece. On the Western margin however compression does occur.

A series of in situ stress measurements have been carried out North and West of Thessaloniki, as well as in central Greece and in Marathon near Athens. Outcrops of limestone and marble were selected in regions of limited topography. Boreholes were drilled to depths reaching 4 meters. Strain gauges ("doorstoppers") were stuck on the bottom of the borehole, and their deformation after overcoring was used to determine the horizontal stress field. This requires a measurement of the elastic moduli of the cores which has not yet been performed. Using standard values for Poisson's ratio and for Young's modulus and the measured deformations, one finds a major axis of extension trending N−S in Northern and Central Greece. In Marathon one single site was investigated and yields NW−SE extension. At a right angle with the above mentioned trend the stress is found to be compressive, i. e. E−W compression, for central and Northern Greece. In Marathon the second major axis is extensive.

These results, as presented in Fig. 1, show that the measured stress field is remarkably homogeneous. They confirm the azimuths of the principal stress axis deduced from neotectonic studies in relationship with recent seismic motions.

Reference

Mercier, J. L.: The Thessaloniki Earthquake. Nature 278, 45—48 (1979).

Address of authors: C. Froidevaux and C. Paquin, Laboratoire de Géophysique et Géodynamique interne, Université Paris Sud, Orsay F-91405, France. C. Angelidis and A. Tzitziras, Institut de Recherches Géologiques et Minières, Départment Géotechnique, 70 rue Messoghion, Athènes 608, Grèce.

Rock Mechanics, Suppl. 9, 21—29 (1980)

Rock Mechanics
Felsmechanik
Mécanique des Roches
© by Springer-Verlag 1980

In Situ Stress Measurements in Austria

By

F. Kohlbeck, K.-H. Roch, and A. E. Scheidegger

With 4 Figures

Abstract

In situ stress measurements were made at 6 locations in Austria, which were chosen in such a fashion as to yield information on the neotectonic stress in the country. Because of a strong influence of the Alpine topography, it was necessary to apply finite-element calculations for the interpretation of the field results. At three locations, significant tectonic stresses were found which could not be explained by gravitational effects alone. The orientation of the tectonic stresses is in conformity with that found by other methods (joint orientations, earthquakes) in Europe (greatest compression NW − SE).

1. General Remarks

The Institute of Geophysics of the Technical University of Vienna has been investigating into the stress field in Austria for several years. In this connection, in situ measurements have been made at 6 different locations. All these are in the Eastern Alps in different geological formations and depths (c. f. Fig. 1, Table 1). The interpretation of the results of the stress measurements is difficult because of the topography of the mountains and because of the complicated nappe-structure of the Alps at a depth which is not very well known even to this day.

In contrast to areas with small differences in height and uniform geological structure, as is the case e. g. in the Canadian Shield, it is necessary in the Alps to take large regional and local effects into account, in order to arrive at the regional stress field. In particular, one cannot compute the overburden pressure simply from the thickness of the overlying rock; rather, one has to take the influence of the surrounding mountains into account. Furthermore, if different (geological) formations are found in the immediate vicinity of the measurements, the elastic constants of the latter must be known. It is then possible to estimate the influence of the surrounding mountains on the measurement results; the corresponding calculations are based on finite-element methods.

The choice of the location of the measurements is particularly important. Near-surface regions can be detached from the underlying rocks or lie in a

0080-3375/80/Suppl. 9/0021/$ 01.80

weathering zone so that they do not reflect the tectonic forces. Slide surfaces above the valley floor can have the effect that the overlying areas are free of tectonic stresses. In mines, one has to take the possibility into account that the excavation and stoping of mine-areas may substantially disturb the virgin stress field.

Table 1. In Situ Stresses and Related Data

Location No.	Sea level Depth (m) Rock	S_h (MPa) S_v	$\dfrac{S_h}{S_v}$	E (GPa) v (—)	S_1 S_2 (MPa) S_3	Direction of S_1 ϕ/θ
Mühlbach, Hk. 1	663 750 greywacke	25 34	0.7	40 0.35	39 ± 3 32 ± 3 13 ± 3	160/70
Felbertal 2	1170 70 schist	14 18	0.8	20—40 0.05—0.15	25 ± 2 16 ± 1 5 ± 2	115/54
Ebriachklamm 3	580 10 diabas	7 3	2	70—90 0.2	9 ± 6 6 ± 3 2 ± 3	*/20
Gleinalmtunnel 4	800 800 gneiss- micaschist	20 24	0.85	40—80 0.02—0.1	32 ± 2 20 ± 4 13 ± 2	162/45
Fohnsdorf 5	—414 1100 sandstone	33 28	1.4	20—40 0.05—0.15	54 ± ? 25 ± ? 11 ± ?	188/12
Bleiberg- "Antoni" 6 a	339 561 limestone (metamorphic)	40 33	1.2	30—60 0.1—0.3	46 ± 3 41 ± 4 27 ± 3	*/35
Bleiberg- "Rudolf" 6 b	674 236 limestone (metamorphic)	27 19	1.4	40—50 0.1—0.2	41 ± 12 21 ± 12 7 ± 6	310/20

S_h .... medium horizontal stress
S_v .... vertical stress
S_1, S_2, S_3 ... greatest, medium and lowest principal stress
ϕ .... Azimuth from N to E
θ .... down dip from horizontal
* .... indicates that maximum and medium principal stresses cannot be distinguished within limits of error
E .... moduls of elasticity
v .... poisson's ratio

The measurements themselves show much scatter largely because of the inhomogeneity of the rock. Thus, error limits of 30% have to be expected in the measured stress components. The corrections for the overburden and topography can lead to even greater errors. In bad cases, particularly in shallow measurement locations and in the immediate civinity of a high mountain range, no statements regarding the regional stress can be made.

Based on the experiences gained from measurement-locations No. 1 and 2 (see below), the subsequent locations were chosen in a topographically symmetrical position, i. e. below the center of a valley or below a mountain

Fig. 1. Locations of measurements and directions of maximum compression in Austria

crest. In this case, it can be expected that the forces resulting from the overburden yield a vertically oriented principal stress direction. This leads to a substantial simplification in the interpretation of the measurement results.

2. Procedure

The in-situ stress measurements were made by means of the door-stopper procedure and the triaxial-cell procedure of Leeman (1971), using CSIR (South Africa) cells and equipment. In both procedures, one cuts rock cores out of the virgin rock and measures their deformation upon destressing. After an independent determination of the elastic parameters of the cores, the rock-stress state can be calculated. In the doorstopper-procedure, one measures the plane strain state at the bore-hole end in 3 bore-holes with differing directions of their axes. In the triaxial-cell procedure, the plane strain state is measured at the interior face of a rock cylinder in a single borehole.

As noted above, it is necessary to determine the elastic constants of the cores for a determination of the rock *stress* state. The corresponding values were determined in uniaxial loading-experiments; for the interpretation of the doorstopper measurements, radial-loading experiments (after Leeman, 1971) were also made. It turned out that the rocks were, in part, substantially anisotropic (Young's moduli in perpendicular directions could vary as 1 : 2) and inhomogeneous (variations to 100%). On the other hand, the load-

dependence of the elastic moduli was very small (a few per cent). The calculation of the stresses from the strains obtained from doorstopper measurements was effected by the computer program of Kohlbeck et al. (1979) by
minimizing the mean squares of the errors of εE, where ε is the measured
strain and E the elastic (Young's) modulus determined by radial loading
experiments with the same axis. It turned out that the results could not be
improved by accounting for the E-modulus individually for each core from
one and the same borehole; it was quite sufficient to take a mean E-modulus
for each borehole. In contrast to our experience, Ribacchi (1977) obtained
a slight improvement by corresponding calculations when the anisotropy of
the rocks in a borehole was taken into account.

3. The Measurements

Location 1

The measurements were made in the former copper mine near Mühl-
bach, in the paleozoic base of the Northern Calcareous Alps, at an elevation
of 663 m above sea level, below an overburden of 750 m thickness. Measurements were made in 3 boreholes by the doorstopper method (cf. Brückl
et al., 1975). The joint evaluation of the results yielded no significant direction for the largest principal stress direction. On the other hand, from the
principal stress directions for each borehole, a most probable maximum
pressure direction was determined as ESE/70 (Azimuth and plunge downward). Thus, the stress state could be explained by the topography.

Location 2

The stress tensor was also determined by the doorstopper method in a
near-surface test-drive of the tungsten mine in the Felber Valley (Carniel
et al., 1976). For this location in the schist-cover of the Central Alps, one
obtained the result that the stress tensor could also be explained by the
topography as well as by the elasto-plastic properties of the rocks (Kohlbeck et al., 1979).

Location 3

Doorstopper measurements were made on 2 boreholes in the Ebriach
gorge, in a narrow vertical diabase dike in the vicinity of the Periadriatic
lineament. In spite of large measurement-errors, significant horizontal stresses
could be detected whose mean value was about 7 MPa. The ratio of mean
horizontal and mean vertical stress is about 2 : 1.

Location 4

A cross-cut was available in the center of the Gleinalm superhighway
tunnel for making stress measurements which made the procedures very
convenient. Doorstopper as well as triaxial-cell measurements were made.
A substantial deflection of the maximum principal stress direction from that

expected from the topography was found. The principal stress direction is, in the mean, SSE/45⁰ (azimuth and plunge downward) and, thus, is normal to the layering. The question whether this orientation can be explained by geological conditions alone is still the subject of investigations.

Location 5

The coal mine near Fohnsdorf was, in 1977, the deepest accessible place in Austria with a depth of 1100 m below ground, 414 m below sea level (the mine has been abandoned and flooded since then). In-situ stress measure-

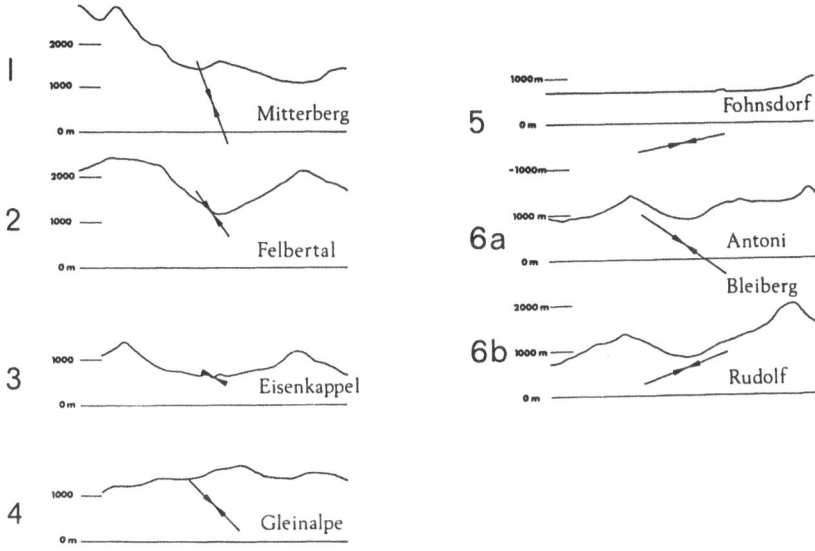

Fig. 2. Directions of principal compressions in vertical sections

ments using doorstoppers and triaxial cells were made. The boreholes were in sandstone. The results indicated very high horizontal stresses in the NS-direction and very small ones in the EW direction. The intermediate principal stress corresponded to the overburden pressure. Since the location was in the center of a broad interalpine depression, it can be assumed that the influence of the topography was very small. The action of a strong N−S directed tectonism is most reasonable at this location.

Locations 6 a and 6 b

These locations were in the lead-zinc mine at Bleiberg in the region of a high valley (900 m above sea level) between the Villach Alpe and the Bleiberger Erzberg. A marked fault striking EW, dipping at 80⁰ to the S runs approximately along the valley floor, from which several, NE striking, steeply dipping minor faults originate.

The measurement-location was 560 m below ground; the doorstopper method was used. It turned out that the smallest principal stress is normal to the layering plunging downward at 55⁰. The intermediate and smallest principal stresses are about equal reading 45 MPa and are thus much larger than the value expected (finite element calculation) from the overburden; this would yield a horizontal stress of 9 MPa.

Location 6b was ca. 2000 m to the East of 6a 236 m below ground; triaxial cells were used. A significant principal stress in the NW direction was found. The stress-values measured in situ were again much higher than those expected from the overburden by finite element calculations.

The results for all locations in plan are shown in Fig. 1 and in cross sections in Fig. 2.

4. Finite-Element Calculations

As has already been mentioned on several occasions, it is necessary to take the topography into account when in-situ stress measurements are to be interpreted. For this reason, calculations of two-dimensional finite-element models were made for the locations No. 1, 2 and 6. Fig. 3 shows the results

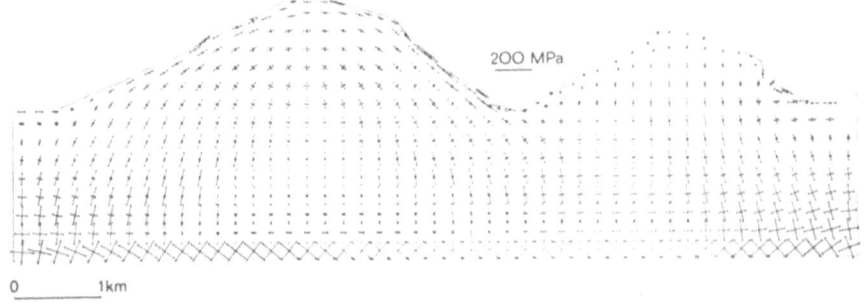

Fig. 3. Finite element results for the measurements in the Felber Valley (Loc. 2)

of such a calculation for Location 2. A purely elastic behavior of the rock was assumed. Since no information of the elastic behavior of the mountain range was available, the elastic constants had to be determined by extensive tests on the cores. The calculations were executed for profiles which were approximately normal to the valley and mountain trends. Because of this, a plane strain state could be assumed. The choice of the boundary conditions had a substantial influence on the result of the calculations. The lateral boundaries were chosen as vertical beneath mountain peak and valley floor. On these boundaries, only vertical displacements were admitted. The lower boundary was taken as a horizontal base. On the example of location 2 (Felber Valley), it was possible to show that a hydrostatic pressure distribution at the base can explain the in-situ values as a consequence of the weight of the overburden. A comparison with the assumption of a rigid base yielded large discrepancies. On the other hand, the in-situ results for the locations 6a and 6b (Bleiberg) could not be explained by the weight of the overburden; neither by the assumption of a hydrostatic pressure distribution at the base nor by a rigid base. Although the horizontal stresses were greater for a hydrostatic pressure distribution at the base than for a rigid base, they do not suffice to explain the high in-situ values. For this, an additional rotation of the mountains or tectonic stresses are necessary.

5. Conclusions

It is known from various, completely independent studies of the present-day tectonic stress-field, such as fault plane solutions of earthquakes (Ahorner, 1975), measurements of horizontal stylolites (Schäfer, 1978), joint orientation measurements (Scheidegger, 1977), tectonic lineaments

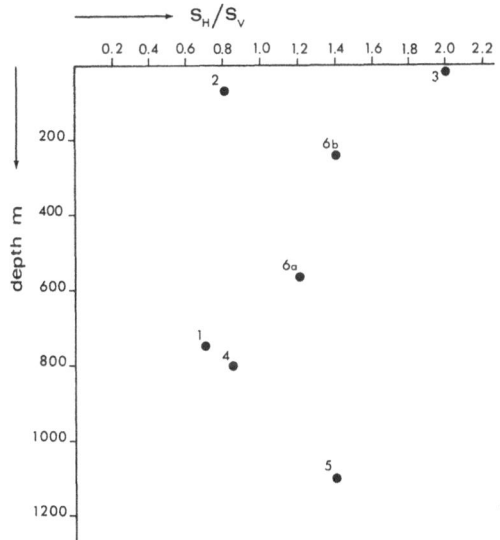

Fig. 4. Value of stress against depth in Austria

(Tollmann, 1977) that a stress system exists in the central European region with a maximum compression oriented in a direction of NW to NNW. In-situ stress measurements in the region of the Rhine Trench (Illies and Greiner, 1978) and upper Italy (Martinetti, 1974) yielded the same result. Whilst 3 of the Austrian in-situ stress measurement locations yielded no certain statements on the neotectonic stress system, the 3 remaining ones yielded a significant maximum principal stress direction oriented NW to N. The complete picture of the orientation of the maximum compressive stress orientations is shown in Figs. 1 and 2. The high horizontal stresses could not be explained by the overburden pressure. If the ratio of mean horizontal to vertical stress is plotted against depth below ground (Fig. 4), no uniform picture is obtained. This is in contrast to measurements in South Africa, where in a region in which there is no pronounced tectonic stress, the relation

$$\sigma_H/\sigma_V = 0.48 + 248/\text{depth (m)}$$

exists (Van Heerden, 1976, Garr and Gray, 1978). If the thermicity of the rocks is disregarded, unloading due to erosion alone (Voight, 1966) could explain such stress conditions. Such an explanation is evidently impossible with measurements of the rates of rising of the Alps and with gravity measurements (Steinhauser and Gutdeutsch, 1976) which show that the displacement rates in the Alps cannot be solely of isostatic origin.

Acknowledgements

The work reported here is the result of investigations carried out as an Austrian contribution to the International Geodynamics Program and was thus financially supported by the Austrian Academy of Sciences. Without this support, the work could not have been done.

References

Ahorner, L.: Present-day Stress Field and Seismotectonic Block Movements Along Major Fault Zones in Central Europe. Tectonophys. 29, 233—249, 251—264 (1975).

Brückl, E., Roch, K.-H., Scheidegger, A. E.: Significance of Stress Measurements in the Hochkönig Massif in Austria. Tectonophysics 29, 315—322 (1975).

Carniel, P., Roch, K.-H.: In-situ-Gebirgsspannungsmessungen in Felbertal, Österreich. Riv. Ital. Geofisica 3, 233—240 (1976).

Garr, A., Gay, N.: State of Stress in the Earth Crust. Ann. Rev. Earth Planet. Sci. 6, 405—436 (1978).

Illies, J. H., Greiner, G.: Rhine Graben and Alpine System. Bull. Geol. Soc. Amer. 89, 770—782 (1978).

Kohlbeck, F., Riehl, G., Roch, K.-H., Scheidegger, A.: In-situ-Spannungsmessungen an der Periadriatischen Naht in der Ebriachklamm bei Eisenkappel. Mitt. Ges. Geol. Bergbaustud. Österr. 26, in press (1979).

Kohlbeck, F., Scheidegger, A., Sturgul, J.: Geomechanical Model of an Alpine Valley. Rock Mech. 12, 1—14 (1979).

Kohlbeck, F.: In-situ-Spannungsmessungen im Tertiärbecken von Fohnsdorf. Berg- u. Hüttenmänn. Monatshefte 124 (8), 367—375 (1980).

Leeman, E. R.: The CSIR "Doorstopper" and Triaxial Rock Stress Measuring Instruments. Rock Mech. 3, 25—50 (1971).

Martinetti, S., Ribacchi, R.: Result of State of Stress Measurements in Different Types of Rock Masses, Proc. 3rd Congress Int. Soc. Rock Mechan. Denver 1974, 2 A, 458—463 (1974).

Ribacchi, R.: Rock Stress Measurements in Anisotropic Rock Masses. Field Measurements in Rock Mechanics. International Symposium Switzerland, April 4—6, 1977, Federal Inst. of Technology, Zurich (1977).

Schäfer, K. H.: Geodynamik an Europas Plattengrenzen. Fridericiana 23, 30—46 (1978).

Scheidegger, A. E.: Geotectonic Stress Determinations in Austria. Proc. Int. Sympos. Field Measurements in Rock Mechanics, Zurich (publ. Balkema, Rotterdam) 1, 197—208 (1977).

Steinhauser, P., Gotdeutsch, R.: Rezente Krustenbewegungen und Isostasie in den Hohen Tauern. Arch. Met. Geophys. Biokl. Ser. A 25, 141—149 (1976).

Tollmann, A.: Faulting Tectonics of Austria on Landsat Photos. N. Jb. Geol. Paläont. Abh. 153, 1—27 (1977).

Van Heerden, W. L.: Practical Application of the CSIR Triaxial Strain Cell for Rock Stress Measurements. Proc. of the Symposium on Exploration for Rock Engineering, Johannisburg, Ed. by Z. T. Bieniawski 1977, A. A. Balkema, Rotterdam (1976).

Voight, B.: Beziehungen zwischen großen horizontalen Spannungen im Gebirge und der Tektonik und der Abtragung. Proc. 1st Congr. Int. Soc. Rock Mech., Lisbon 1966, 2, 51—56 (1966).

Address of authors: Dr. F. Kohlbeck, Dr. K.-H. Roch, Prof. Dr. A. E. Scheidegger, Institute of Geophysics, Technical University, Gusshausstrasse 27—29, A-1040 Wien, Austria.

Rock Mechanics, Suppl. 9, 31—47 (1980)

Rock Mechanics
Felsmechanik
Mécanique des Roches
© by Springer-Verlag 1980

In Situ Stress Measurements in Italy

By

Sandro Martinetti and Renato Ribacchi

With 9 Figures

Abstract

Most of the in situ natural state-of-stress measurements have been carried out in Italy at the sites of large underground hydroelectric powerhouses and at the sites of important mines. The method successfully used was in all cases based on the measurement of the deformations caused by the stress release on the bottom of boreholes. Some of the rock masses in which measurements were taken, showed a considerable anisotropic behavior; for these cases the interpretation of the results provided by the measurements appeared to be more complex and not so reliable. A method of approximate interpretation has been adopted in order to overcome such difficulties.

An overall assessment of the results seems to indicate that the measured state of stress is strongly influenced by the morphological situation and it is not easily related to the regional tectonic situation; in fact, the measurements were often taken at shallow depths and in localities having rather uneven morphology (for example on narrow valley beds or on very steep slopes). In some cases the values of the horizontal component of the state of stress along the axis of the valleys appeared to be quite high; high horizontal stresses were measured also in the Raibl mine which is subject to considerable rock bursts.

1. Introduction

Complex procedures and considerable costs are involved in the experimental determination of the original state of stress in rock masses. These measures are thus mainly carried out only for important civil engineering works in limited areas and for important mining exploitations.

In Italy, over the last ten years, the original state of stress was measured in the underground powerhouses of the ENEL (Italian State Electricity Board) hydroelectric plants and in some important mines (Fig. 1). The method adopted is known as the CSIR "doorstopper" method (Leeman, 1969); the strains caused by the stress release due to "overcoring" are measured on the flattened bottom of a borehole.

Most of the measures were obtained directly by the authors; for the sake of completeness the paper also includes the measurements performed

0080-3375/80/Suppl. 9/0031/$ 03.40

in the mines of Raibl and S. Giovanni, whose data have been taken from the literature (Stragiotti et al., 1976; Cotza et al., 1974). The results of the measurements carried out in the underground power plants, performed by the Geotechnical Service of ENEL, have already been partially published (Martinetti and Ribacchi, 1974; Martinetti, 1977); however, on this

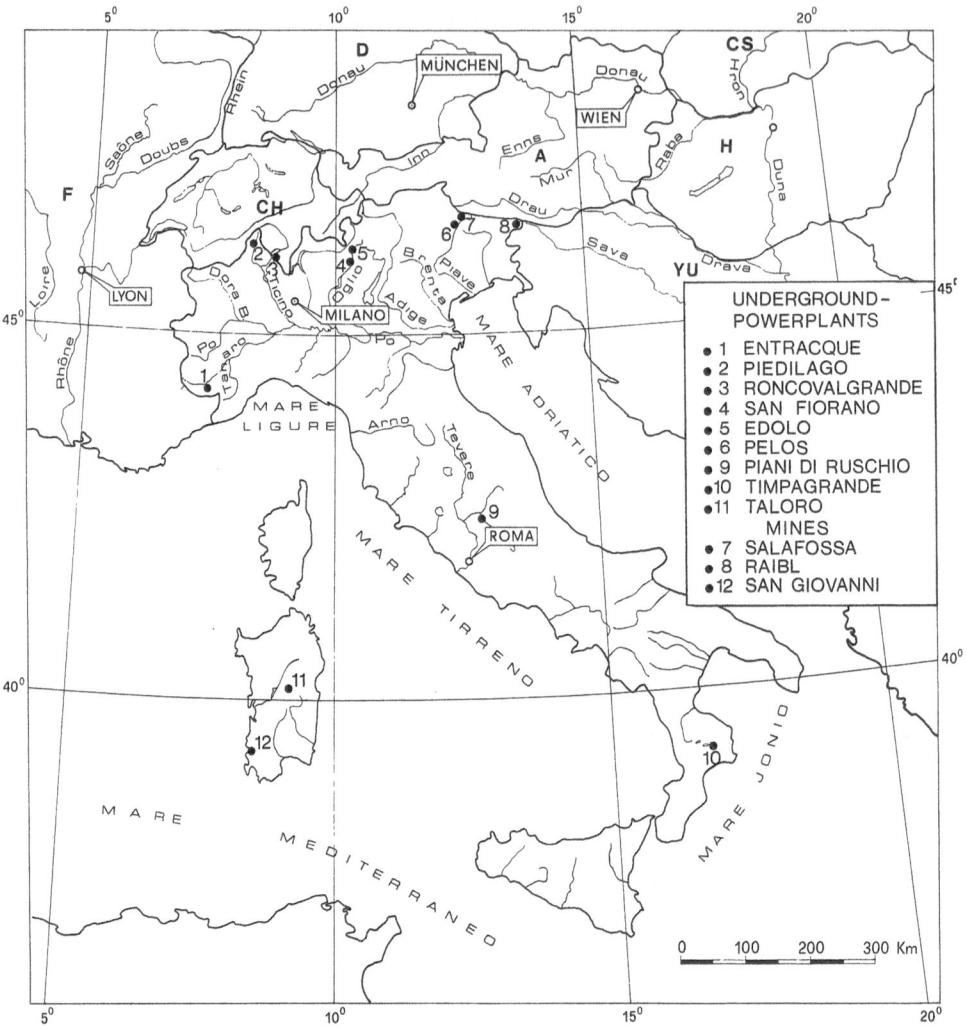

Fig. 1. Location of the measurement sites

occasion all the available data were reprocessed taking into account a more accurate assessment of the elastic characteristics of the rock and also considering its anisotropy, if any.

2. Remarks on the Measurement Method

The reasons for which the CSIR "doorstopper" method is practically one of the most convenient have already been extensively discussed previously (Martinetti and Ribacchi, 1974). The most meaningful advantage of this method lies in the fact that the measurements can be carried out also in highly fractured rocks where it is practically impossible to obtain cores measuring a few decimetres as required for other methods (for example the triaxial CSIR, Leeman, 1969).

The CSIR "doorstopper" method may seem more complex than the methods of the triaxial CSIR type. In fact, the former implies, among other things, that the measurements be taken in at least three boreholes having different orientations and also that an overall interpretation be given on the basis of the results obtained.

On the contrary, for example, with the triaxial CSIR method the state of stress could theoretically be determined by means of only one measurement in only one borehole. However, practice has proved that the stress distribution in a rock mass varies from one point to another because of the lithologic and structural non-homogeneities; it is therefore, not possible to obtain a correct evaluation of the "global" original state of stress without carrying out "point" measurements in such a number and positions as to allow a meaningful statistical analysis of the results. These factors, by far, reduce the theoretical advantages of the methods of the CSIR triaxial type, even when they can be applied in practice.

The results provided by the CSIR "doorstopper" method are interpreted by means of a linear regression analysis, the regressors being the six components of the original state of stress (Martinetti and Ribacchi, 1970). Through this statistical model it was possible for us to evaluate also the influence on the accuracy of the results, of the mutual orientation of the measurement boreholes and of the orientation of the strain gauges in each borehole (Martinetti and Ribacchi, 1970). Thus a series of configurations have been outlined, which are "convenient" from a theoretical standpoint and which are technically feasible taking into account the limitations imposed by field requirements. Experience has shown also that it is convenient to use four measurement boreholes, since this allows to check the validity of the results by comparing the interpretations obtained from the four combinations of the holes considered three at a time. On an average 8 three-strain gauge cells are used in each borehole, totalling about 100 strain measurements.

For the interpretation of the results, the stress concentration factors that link the original state of stress to the state of stress on the flattened bottom of the hole are required. They are known with good accuracy only for isotropic rocks; instead, for anisotropic rocks, the coefficients can be calculated only for one case at a time by means of complex and expensive procedures; however, satisfactory results can be obtained through an approximate method which is described in detail in a previous paper, Ribacchi (1977).

This method requires a uniform radial pressure to be applied to the lateral surface of the cores (obtained through overcoring) to which the strain cells are stuck. For anisotropic rocks the deformations measured by the strain gauge of the cell are connected by a sine law to the orientation of the strain gauges themselves with respect to the fabric of the rock (Fig. 2).

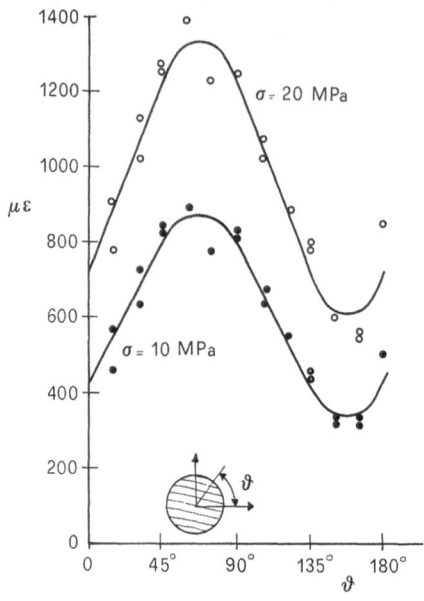

Fig. 2. Strains measured by the "doorstopper" cells on the cores from one of the measurement holes at the Piedilago site, following application of a radial pressure on the lateral surface of the core

In the interpretation of the results, the in situ strain measured by each gauge (because of overcoring) is corrected by scaling it proportionally to the strain induced in the gauge itself by applying a fixed radial pressure on the cores; by using such corrected values the original state of stress is then evaluated by considering the rock as being isotropic. In practice, the procedure utilizes the same coefficients of stress concentration which are valid for an isotropic rock (and this is obviously an approximation) and it keeps account of the anisotropy only for the deformability variations of the rock with the direction.

The radial pressure tests on cores obtained through overcoring were carried out also for isotropic rocks; thus a direct assessment of the elastic modulus of the rocks was obtained, which was used for the global interpretation of the data. This type of test is certainly more significant than the traditional uniaxial compression tests, because the stress conditions are more similar to those existing on the bottom of the borehole. Poisson's coefficients, instead, were obtained by means of uniaxial compression tests.

Quite often the tested rocks showed a non-linear behaviour, with moduli increasing with the applied load; this behaviour, due to the gradual closing of the microfissures, is particularly apparent in the anisotropic rocks and it could bias the interpretation of the stress measurements. In order to achieve greater accuracy, it was decided to adopt the rock deformability values at stress levels corresponding to those existing in the measurement sites.

The foregoing shows that for most cases the principal directions of the original stress tensor, that are not affected by inaccuracies in the assessment of the elastic moduli of the rock, are determined with greater reliability than the absolute values of the principal stresses.

3. Examination of Results

The plans in Fig. 3 and the cross-sections in Fig. 4 show the morphology of the areas where the measurements were performed.

In most cases the measurements were carried out in valley slopes having great longitudinal extension and, in a few instances, in partially isolated mountain buttresses. The slopes where the measurements were performed have a considerable height, mostly over 1000 m. As a consequence, the measurement sites must be considered relatively "superficial" with respect to the slopes themselves (Fig. 4).

Table 1 briefly illustrates the geological setting and the morphology of the various sites; it also provides the values of the mechanical parameters adopted for the interpretation of the results.

The values of the principal stresses and their orientations are shown in Fig. 5. The confidence limits of the principal directions in each principal plane are also indicated. It must be noted that a considerable extension of these limits does not necessarily mean that the measurements are highly scattered, but could simply mean that in that particular plane the original state of stress is almost isotropic. In the same figure we indicated, when available, the contour lines of the fracture intensity calculated by a computer program directly on the spherical reference surface of the stereographic projection.

Figs. 3 and 4 show the directions and the values of the secondary principal stresses respectively in the horizontal plane and in the plane of the vertical section.

A close examination of the results leads to the following remarks:

— In mountain slopes or buttresses, the inclination of the maximum stress in the vertical section falls between that of the slope and the vertical (Fig. 6). This is in agreement with the theoretical evaluations of the stresses that are generated within the rock mass following the excavation of the valley. Fig. 7 shows an example of the results of this type of evaluation carried out by using finite element techniques and by assuming an elastic behaviour of the rock mass. The values of the state of stress resulting from this type of analysis depend upon assumptions regarding the state of stress existing before the valley was formed (and particularly the intensity of the horizontal component); however, within the range of reasonable assumptions that can be made, the effect of possible horizontal extragravitational components is not very important.

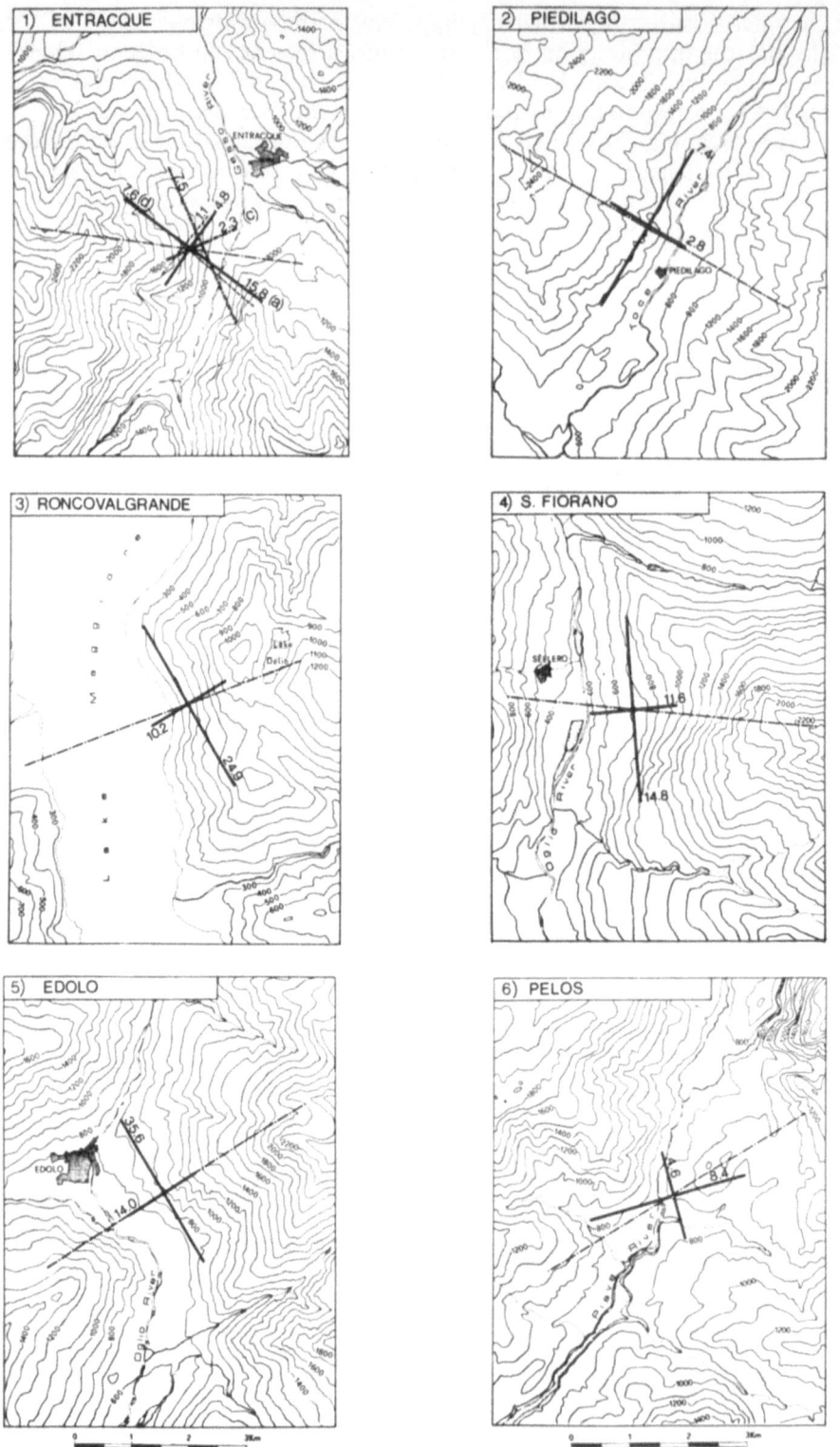

Fig. 3

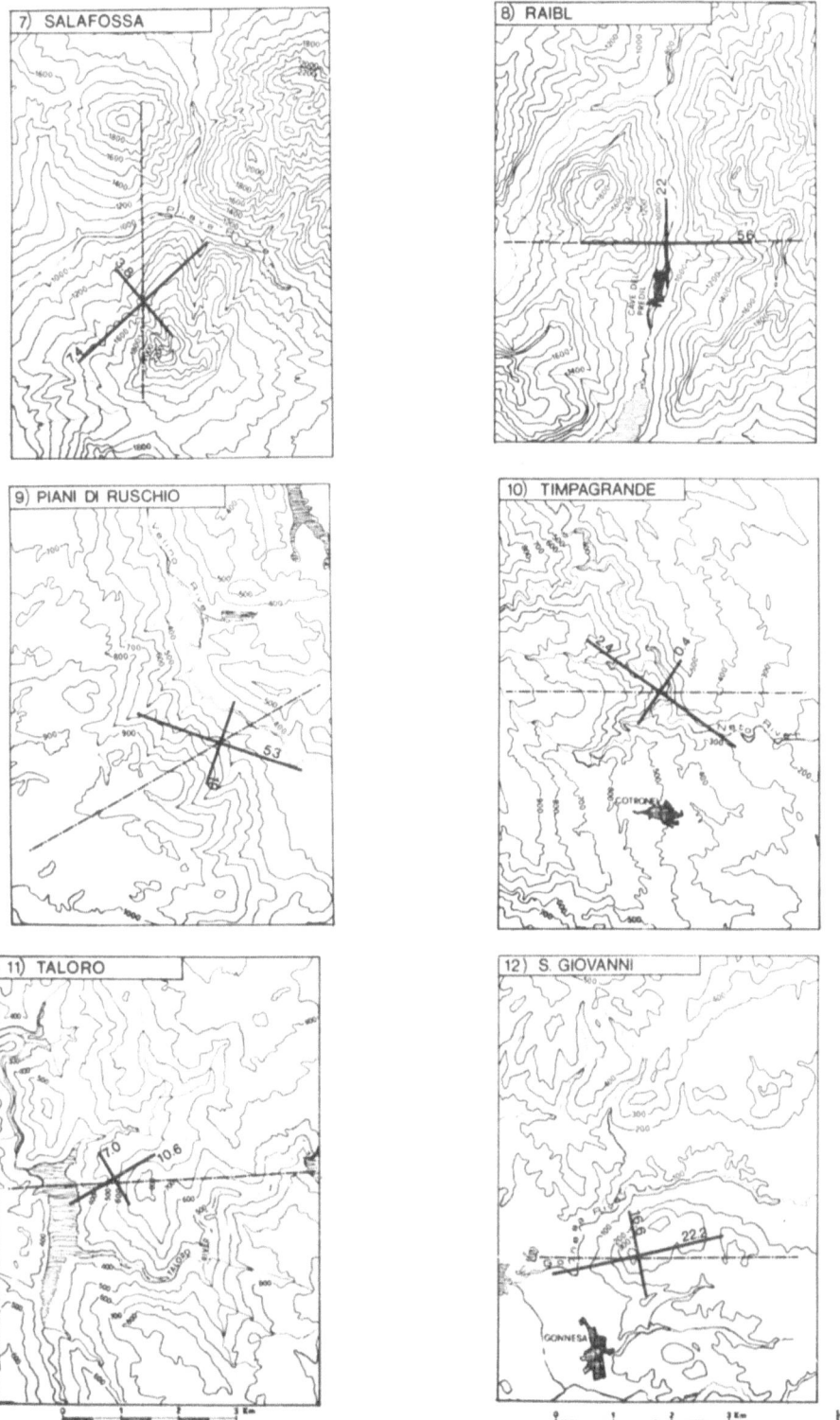

Fig. 3. Plans of the investigated sites and secondary principal stresses in the horizontal plane. The cross-section paths are indicated

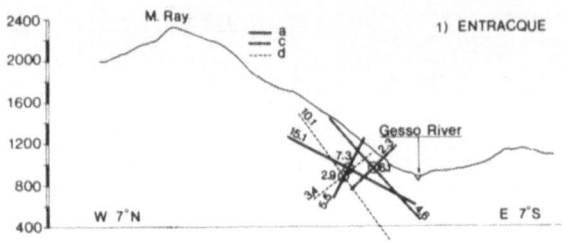

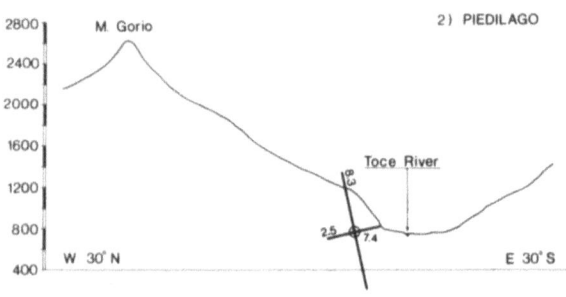

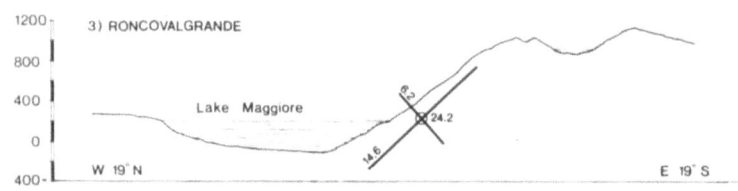

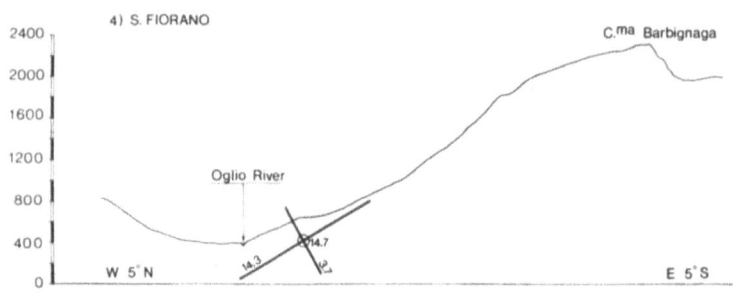

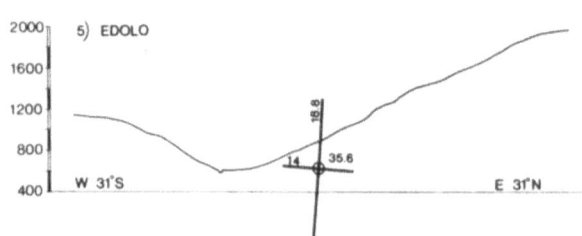

a

Fig. 4

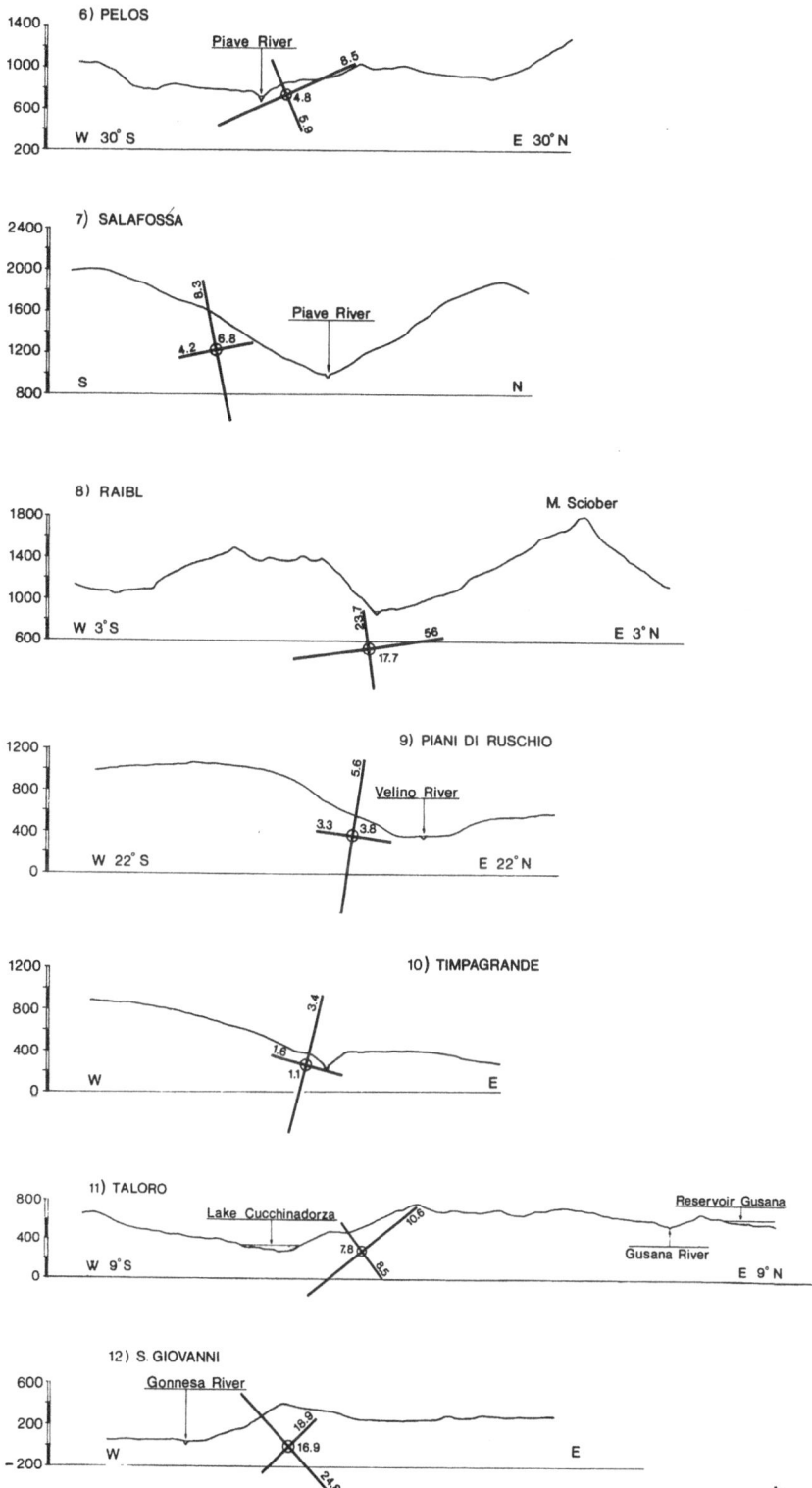

Fig. 4. Vertical cross-sections of the measurement sites and secondary principal stresses in the same plane

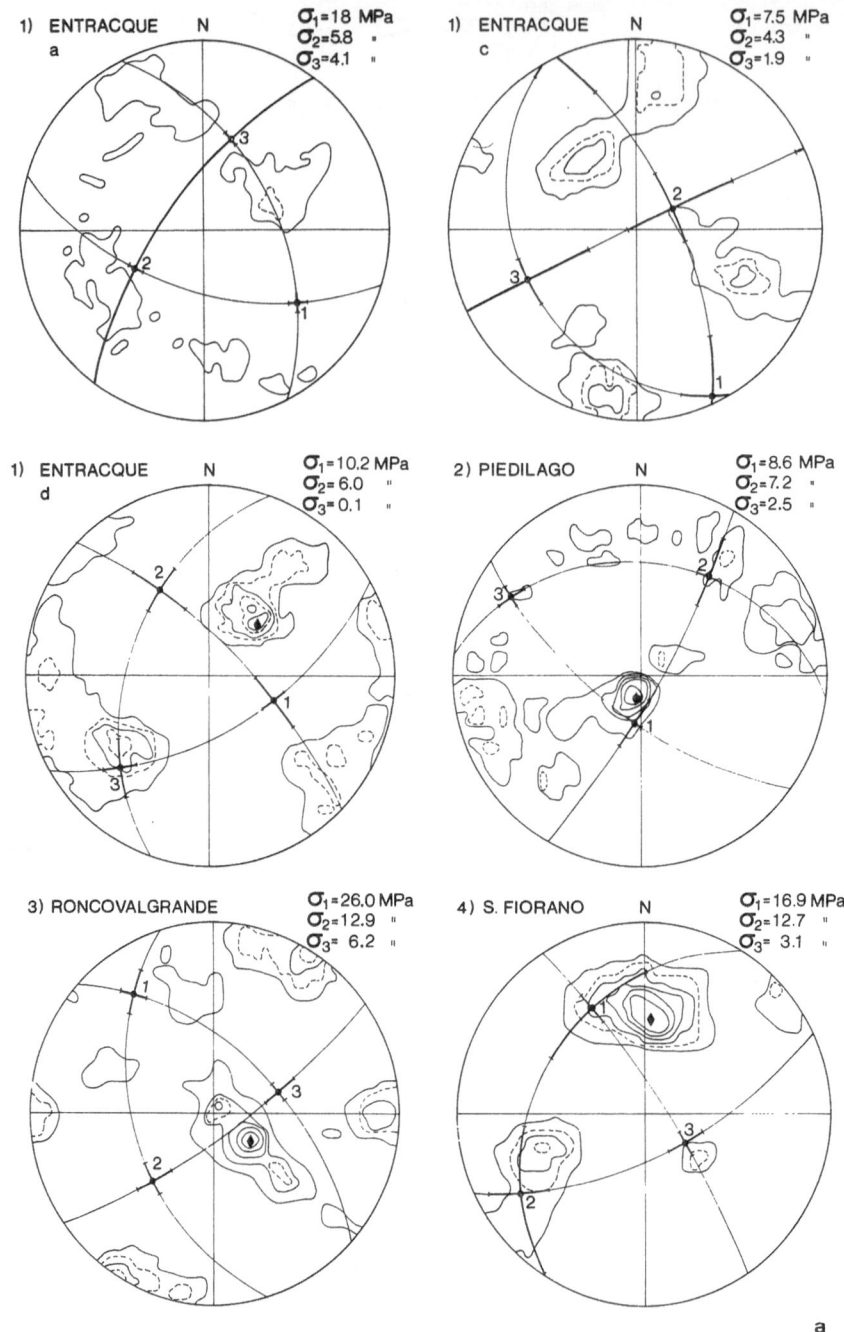

Fig. 5

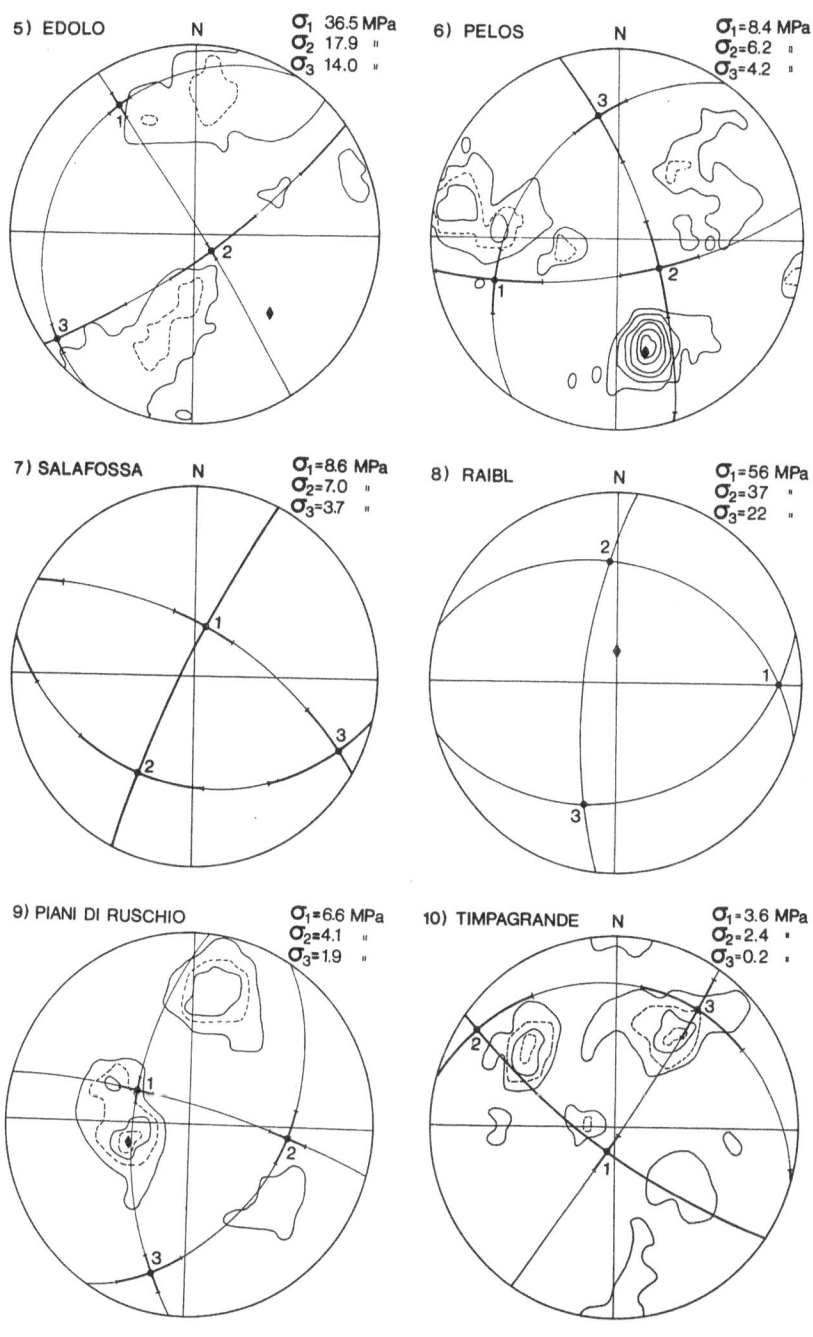

Fig. 5

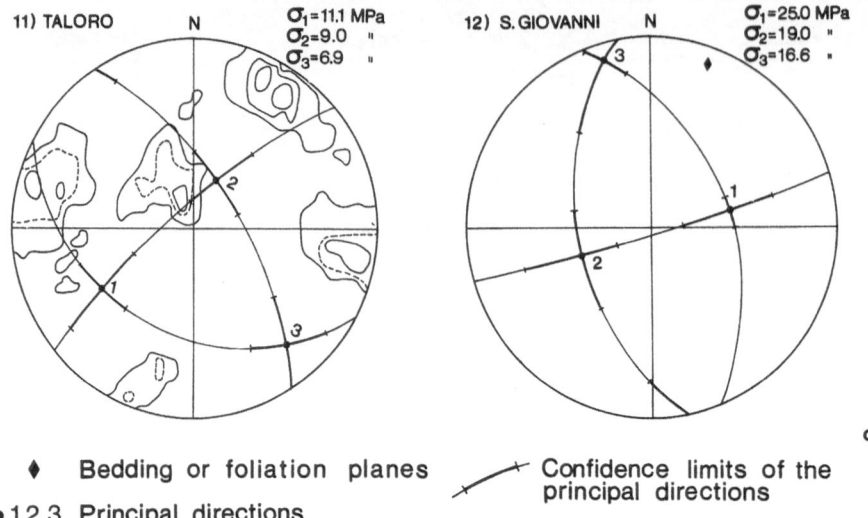

11) TALORO   N $\sigma_1 = 11.1$ MPa
                               $\sigma_2 = 9.0$ "
                               $\sigma_3 = 6.9$ "

12) S. GIOVANNI   N $\sigma_1 = 25.0$ MPa
                                    $\sigma_2 = 19.0$ "
                                    $\sigma_3 = 16.6$ "

◆ **Bedding or foliation planes**   ┼ **Confidence limits of the**
                                              principal directions

● **1,2,3 Principal directions**

Fig. 5. Principal values of the state of stress and contour lines of fracture intensity (Wulff stereographic projection, lower hemisphere). The interval of the contour lines is 2% (full lines) or 1% (dashed lines)

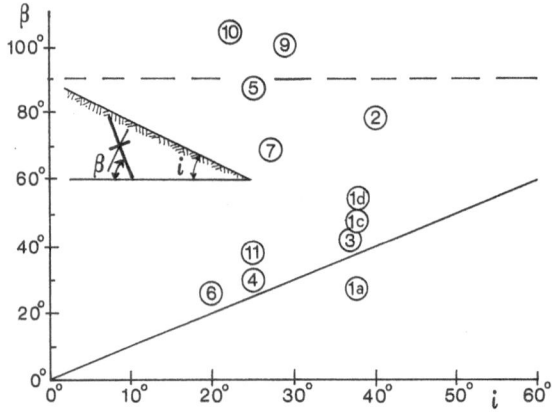

Fig. 6. Inclination β of the greater secondary stress versus the slope inclination i

— The vertical stresses σ_V are mostly equal or greater than the γh values that can be calculated on the basis of the overburden thickness h (Fig. 8). Theoretical analyses carried out by assuming either elastic or elasto-plastic behaviour of the rocks show that in correspondence to a slope, σ_V should be somewhat higher than γh. This does not occur in our case only for the Entracque and Piedilago sites; quite probably for these sites we underestimated the elastic moduli of the rock used for processing the experimental results.

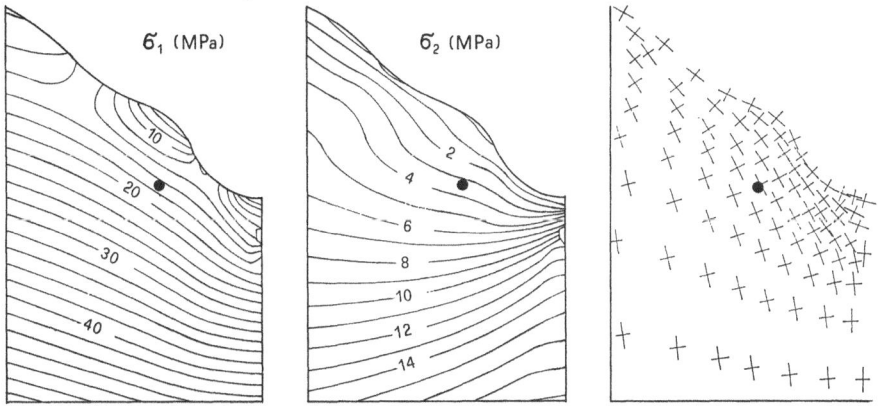

Fig. 7. State of stress at the Piedilago site evaluated by means of a finite element elastic analysis. The state of stress before the excavation of the valley is assumed to be purely gravitational

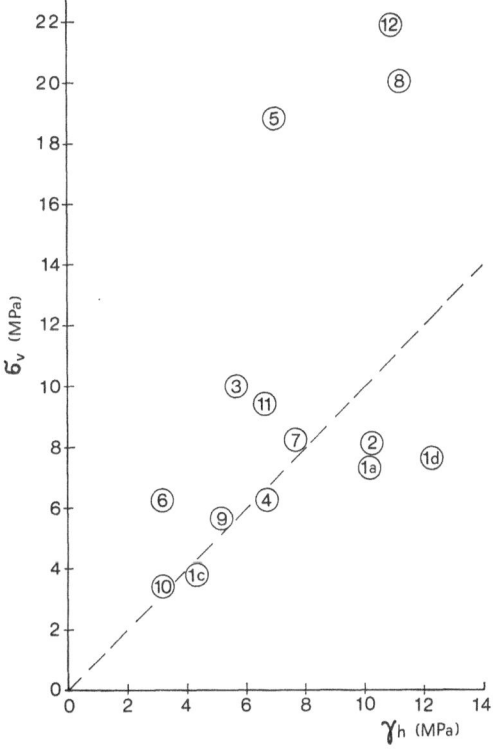

Fig. 8. Vertical stress σ_v in the test sites versus overburden pressure γh

— In the Raibl and in the S. Giovanni mines, the vertical stresses are by far greater than those corresponding to the overburden. For the former, this is certainly due to the fact that the measuring site is located on

Table 1. Characteristics of the Measurement Sites

Site	Rock type and formation	Fracture system	Morphology	Mechanical parameters
1 Entracque a	Anatexites of the Argentera massif composed of quartz and feldspar with low mica content	Not very systematic $n = 1.5 \text{ m}^{-1}$	Mountain slope $i = 38°$, $H = 1500$ m, $h = 380$ m	$E_1/E_3 < 1.5$ $\bar{E} = 48.5$ GPa $\nu = 0.07$ $\gamma = 26.9 \text{ kN/m}^3$
Entracque c	— Same —	— Same —	$h = 160$ m	— Same —
Entracque d	Belt of paleomilonite included in the anatexites	Foliation joints and 2 joint sets $n = 4.7 \text{ m}^{-1}$	$h = 480$ m	Anisotropic $E_1/E_3 = 2.0$ $\bar{E} = 31$ GPa $\gamma = 26.5 \text{ kN/m}^3$
2 Piedilago	Gneiss (20% quartz, 50% feldspar, 20% mica). Antigorio formation of the "Piemontese" zone	Subhorizontal foliation joints and a girdle of joints almost normal to foliation $n = 1.5 \text{ m}^{-1}$. Two main fault sets in the area have E 10° N and E 40° S strikes	Mountain slope on the Eastern side of Toce valley. $i = 40°$, $H = 1600$ m. The slope is steeper at low elevations above the river bed. $h = 390$ m	Anisotropic $E_1/E_3 = 2.5$ $\bar{E} = 15.5$ GPa $\gamma = 26.5 \text{ kN/m}^3$
3 Roncoval-grande	Fine-grained gneiss having low mica content. "Scisti dei Laghi formation" of the Southern Alps crystalline basement	3 joint sets $n = 5 \text{ m}^{-1}$	Mountain slope $i = 37°$, $H = 1000$ m, $h = 210$ m	$\bar{E} = 25$ GPa $\nu = 0.05$ $\gamma = 27.3 \text{ kN/m}^3$
4 S. Fiorano	Phyllite (50% quartz and feldspar, 50% mica and chlorite). "Scisti di Edolo" formation of the Southern Alps crystalline basement	Foliation joints and 2 joint sets $n = 1.7 \text{ m}^{-1}$	Mountain slope on the left side of Camonica valley $i = 25°$, $H = 1800$ m, $h = 240$ m	Anisotropic $E_1/E_3 = 2.0$ $\bar{E} = 12$ GPa $\gamma = 27.9 \text{ kN/m}^3$
5 Edolo	— Same —	Foliation joints less marked than in 4. Tectonic lines (Tonale line and Gallinera line) have locally a NNE or NE strike (parallel to foliation)	Mountain slope on leftside of Camonica valley $i = 25°$, $H = 1700$ m, $h = 250$ m	Anisotropic $E_1/E_3 = 4.5$ $\bar{E} = 22.5$ GPa $\gamma = 27.9 \text{ kN/m}^3$

Site	Rock type and formation	Fracture System	Morphology	Mechanical parameters
6 Pelos	Thinly-bedded limestones with marly interlayers. Werfen formation (Lower Trias Age)	Bedding joints and at least 3 other sets	Low elevation hill between the Piave River and the limestone mountains at the back. h = 120 m	$\bar{E}$ = 56 GPa ν = 0.23 γ = 26.5 kN/m³
7 Salafossa	Dolomitic limestone of the "Dolomia Metallifera" formation (Middle Trias Age)	Relatively intense	Mountain buttress on the Southern side of Piave valley $i = 27°$, $H = 1200$ m, $h = 280$ m	$\bar{E}$ = 26.5 GPa ν = 0.16 γ = 27.5 kN/m³
8 Raibl	Dolomite of the "Dolomia Metallifera" formation (Middle Trias Age)	Relatively intense. Two main fault sets having a N-S strike and a NE-SW strike	Measurement site is located under the bottom of a N-S trending valley bordered by high mountains. h = 400 m	$\bar{E}$ = 56 GPa ν = 0.25 γ = 28.0 kN/m³
9 Piani di Ruschio	Bedded limestone of the "Maiolica" formation (Lower Cretaceous Age)	Bedding joints and two other sets perpendicular to the bedding, $n = 9$ m^{-1}. Regional tectonic structure characterized by NW and NE striking faults	Mountain slope $i = 29°$, $H = 750$ m, $h = 100$ m	$\bar{E}$ = 39 GPa ν = 0.26 γ = 26.0 kN/m³
10 Timpagrande	Granite (30% quartz, 50% feldspar, 20% mica) of the Calabride complex	4 main joint sets $i = 9$ m^{-1}	Mountain buttress at the confluence of two rivers $i = 32°$, $H = 300$ m, $h = 120$ m	$\bar{E}$ = 53 GPa ν = 0.16 γ = 26.1 kN/m³
11 Taloro	Fine-grained granodiorites of the Sardo-Corso intrusive complex	4 main joint sets, $n = 8.5$ m^{-1}. The main fault sets have a N 40° W and a W 10° N strike	Plateau deeply cut by the Taloro River and its tributaries $i = 25°$, $H = 550$ m, $h = 240$ m	$\bar{E}$ = 34.5 GPa ν = 0.11 γ = 26.1 kN/m³
12 S. Giovanni	Fine-grained limestone of the "Calcare Ceroide" formation (Cambrian Age)	Not very intensive	Measurement point is located under a hill at lower elevation than the surrounding valleys. h = 400 m	$\bar{E}$ = 82.5 GPa ν = 0.33 γ = 27.5 kN/m³

n = fracture frequency; i = mountain slope angle; H = height of the slope; h = depth below ground surface; $\bar{E}$ = Young modulus; ν = Poisson coefficient; γ = unit weight of the rock.

the floor of a valley which is surrounded by high mountains; as to the latter, it is more difficult to supply an explanation.

— The principal secondary stresses on the horizontal plane are often oriented according to the axis of the valley. In some instances (Roncovalgrande, S. Fiorano, Edolo) the longitudinal component is considerably high; on the other hand, in the Timpagrande site, the deep incision of the two valleys that isolate the buttress which includes the measuring zones, justifies the low values of this component. Finally, the high values of the horizontal stresses in the Raibl mine must be pointed out; they are probably one of the factors that account for the rock bursts occurring in the mine. Similar phenomena occur in the neighbouring Bleiberg mine, in Austria, where the geologic situation is quite similar.

— The values of the ratio between minimum and maximum principal stresses vary within a very wide range; however for most cases they fall within the 0.15—0.40 range (Fig. 9). The lower limit of this ratio is determined by the limit strength of the rock mass, corresponding to the long-term strength of its weaker elements (more highly fractured zones, faults). By assuming a 35^0 friction angle and the absence of cohesion, the value of the ratio could drop to 0.25; the presence of some cohesion could further lower this value.

It must finally be recalled that we have not yet carried out any research on the relationship between the state of stress measured with the "door-stopper" method and the regional state of stress that could be inferred on the basis of the joints distribution according to the methods proposed by

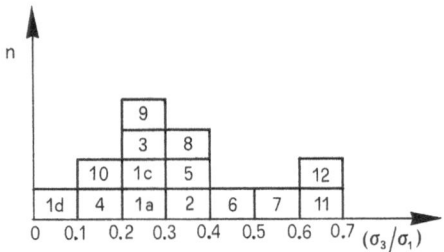

Fig. 9. Histogram of the ratios between minimum and maximum principal stresses for all the testing sites

Scheidegger (1977). The data currently available on the fracture characteristics of the rock masses are however shown in Table 1 and in Fig. 3. It must be pointed out that in our case this type of analysis would certainly incurr two important difficulties. In fact, the state of fracturing appears to be often quite complex and could be due to the superimposition in successive stages of effects of variable stress conditions which could be connected with important morphology modifications. Secondly, the state of stress measured during our work appears to be strongly influenced by the topographic situation which could have deeply modified the "regional" state of stress in the area.

Acknowledgements

The research was partly supported by the CNR contract CT 78.326.05 and by the CNR Centre for Technical Geology.

References

Cotza, R., Siotto, G., Trois, P.: Misure in situ per la determinazione degli sforzi originari in un ammasso roccioso. L'Ind. Min. *25*, 210—219 (1974).

Leeman, E. R.: The C. S. I. R. Doorstopper and Triaxial Rock Stress Measuring Instruments. J. South Afr. Inst. Min. Metall. *69*, 305—339 (1969).

Martinetti, S.: Experience in Field Measurements for Underground Power Stations in Italy. Proc. Int. Symp. "Field Measurements in Rock Mechanics" *2*, 509—534, Zurich (1977).

Martinetti, S., Ribacchi, R.: Un criterio statistico per l'interpretazione dei risultati di misure dello stato di sollecitazione negli ammassi rocciosi. Riv. It. Geotecnica *4*, 21—32 (1970).

Martinetti, S., Ribacchi, R.: Determinazione sperimentale dello stato di sollecitazione originario negli ammassi rocciosi. — Influenza della posizione delle basi di misura sulla precisione dei risultati. X Convegno di Geotecnica, II-10, Bari (1970).

Martinetti, S., Ribacchi, R.: Results of State of Stress Measurements in Different Types of Rock Masses. 3rd Congr. Int. Soc. Rock Mech., Vol. 2 A, 458—463, Denver (1974).

Ribacchi, R.: Rock Stress Measurements in Anisotropic Rock Masses. Proc. Int. Symp. "Field Measurements in Rock Mechanics" *1*, 183—196, Zurich (1977).

Scheidegger, A. E.: Geotectonic Stress-Determinations in Austria. Proc. Int. Symp. "Field Measurements in Rock Mechanics" *1*, 197—208, Zurich (1977).

Stragiotti, L., Armando, E., Pelizza, S.: Indagini geofisiche e geomeccaniche svolte nella miniera di Raibl dall'Istituto di Arte Mineraria del Politecnico di Torino in rapporto al fenomeno del colpo di tensione. Boll. Ass. Min. Subalpina *13*, 544—580 (1976).

Address of authors: Sandro Martinetti, Enel DCo, Geotechnical Service, Rome; Prof. Renato Ribacchi, Institute of Mining, Faculty of Engineering, University of Rome, Roma, Italy.

Rock Mechanics, Suppl. 9, 49—61 (1980)

Rock Mechanics
Felsmechanik
Mécanique des Roches
© by Springer-Verlag 1980

In Situ Strain Measurements in Libya

By

Karlheinz Schäfer

With 1 Figure

Abstract

For the first time, *in situ* rock strain measurements have been carried out in Libya. Except in Liberia there are no previous *in situ* stress results of the African continent north of the equator. In this paper I report the results of stress determinations at 26 sites that extend from the Tunisian-Libyan border to Derna/Cyrenaica in the north and to Ghadames and the Jufrah oasis in the south. The present rock strain has been determined by overcoring of strain gauges that were bonded to the rock surfaces. The range of rock ages selected for measurements was wide (Middle Triassic to Quaternary), but their lithologic character was similar (solid homogeneous micrites). The orientations of maximum and minimum horizontal stress revealed a consistency across large areas. Most stress values vary from 0—50 bars; many range between 50 and 150 bars, and at one site, the stress was tensile in all horizontal directions. At least three crustal domains in northern and central Libya can be defined by means of different *in situ* stress fields. The western Libyan stress domain extends from Tunisia into Libya and is terminated along a line that runs from Sabratah to Azizia, Bu Ngem and the Jabal Waddan east of the Hon graben. This western Libyan stress field has a NW—SE-oriented maximum compressive stress that is horizontal. To the east, the central Libyan stress domain has a NE—SW-directed horizontal maximum compressive stress component that incorporates the entire Sirte basin and major parts of Cyrenaica. A third *in situ* stress domain occurs along the Cyrenaican coastal area from Al Beda to the east. There, the stress field has a maximum compressive component of NW—SE-direction. It is suggested that Tripolitania and the eastern Cyrenaica and their lithospheric northern extensions are indenting the European plate in a northwestern direction corresponding to the drift direction of the African plate for the last 9 m. y. The Sirte basin is located between those framing indenters and was subjected to extensional tectonics during the late Mesozoic and most of the Tertiary, but may have been under NE—SW-directed horizontal compression since the late Neogene.

Introduction

Until a few years ago, *in situ* measurements of rock strain and the determination of rock stress were exclusively confined to problems of civil engineering. Several methods are in use of which hydrofracturing (Haimson and Fairhurst, 1970) and overcoring (Leeman, 1969) are most prominent.

0080-3375/80/Suppl. 9/0049/$ 02.60

Since the concept of plate tectonics became more and more accepted, an increasing number of scientists are discussing the driving mechanism of the new global tectonics. It is out of the question now that the knowledge of the amount of rock stress and its orientation across plate margins or large intraplate areas would decisively contribute to the question as to whether there is push from the rifts, pull from the subduction zones, drag at the plates bases, or a combination of these three major forces. Each of these mechanisms induces stress to the lithospheric plates that is also effective at the surface of the crust.

Besides, there are other processes such as rapid removal or addition of rock overburden, thermal stress by geothermal anomalies, membrane stress by plate movement (Turcotte and Oxburgh, 1976), and changes in crustal thickness from which additional stress may be exerted. Also, the aforementioned stress measuring techniques have been applied mainly in mines, tunnels or deep boreholes to avoid near-surface stress anomalies caused by topography, weathering or daily and seasonal temperature changes.

Nevertheless, in this paper *in situ* stress results are published that have been obtained at the rock surface by the overcoring method, since in Libya it was possible to select sufficient locations for stress measurements where topographic effects and rock weathering were negligible. The results of field tests and laboratory experiments made it possible to eliminate thermally induced stresses in near-surface rocks. Thus, in Libya, measured rock strain has been caused by tectonism rather than by the other above-mentioned stress forming processes.

The results of rock strain measurements in Libya may represent valuable facts for plate tectonic and economic studies. From the plate tectonic viewpoint it is expected that the drifting African lithospheric plate while colliding with the European plate has developed different tectonic regimes and stress domains along its northern margin. From the results of previous structural investigations in Tunisia and Libya (Schäfer, 1978, 1980, this volume), it is assumed that northwestern Libya was part of a tectonic regime that incorporates the northern areas of the Maghreb countries whereas the central and eastern part of northern Libya was contemporaneously under different tectonic and stress conditions than the west. The knowledge of the orientations of principal stress components and the size of Recent stress fields in Libya, as well as the invisaged measurements of residual rock stresses that derive from palaeo-stresses, all contribute to the understanding of this scientific problem. Even more valuable seems to be the knowledge of present-day rock stresses from the economic viewpoint. It is suggested that in the future no critical foundation of a building, no tunneling, mining, or planning of new quarries, no search for fluids in rock cavities should be considered without the knowledge of the orientation of the components of principal rock stresses in that particular area.

If we consider only the migration of fluids through rocks or the location of fluid concentrations in rocks, we must be aware that faults, joints, veins, pores and other possible paths of fluid migration and places of fluid concentration are closed if they strike perpendicular or oblique to the direction of principal compressive stress and are open if they run perpendicular to the main tensile or least compressive stress. The knowledge of pre-existent discontinuities (Schäfer, 1980, this volume) helps to determine what cavities occurring within the rocks (also deep-seated) are open and which are closed.

The Investigated Area

The sites for *in situ* strain measurements have been selected using many geological criteria of which the most important were:

1. Horizontal rock surfaces must be exposed at locations remote from steep and high escarpments or deep canyons.

2. Horizontal rock surfaces of quarry floors must be remote from quarry walls to avoid horizontal stress caused by the deadweight of rocks.

3. Quarries in operation were avoided because of the effect of blasting on the present stress field.

4. The rock surfaces to which the strain gauges were bonded have to be free of joints, cavities and coarse-grained components.

Lithology and Stratigraphy of the Sites

All measurements were carried out on sedimentary rocks. The common occurrence of carbonates in Libya enabled the studies to be concentrated on rocks of comparable lithology. Carbonates with high clay contamination were not considered because of bonding problems. Rather pure and solid micrites and fine-grained dolomites showed the best results. They had a short strain relaxation time after overcoring and their values of Young's modulus were highest among carbonate rocks. Studies of *in situ* strain to determine the orientation of the present-day stress field in various areas in Libya were performed on rocks varying in age from the Azizia Formation of the Middle Triassic to well-cemented Quaternary breccia. The reason for incorporating rocks of such a wide range of stratigraphic ages was to determine whether or not there were residual stresses in older sequences. Discoveries of residual stresses locked in rocks suggest the existence of previous stress fields with orientations of the principal stress components other than the present ones.

Geographic Location of the Sites

As can be seen from Fig. 1, the measurements are concentrated in Tripolitania and in Cyrenaica. The Tripolitanian sites are located in the Jefara plain, on the Jefren-Garian-Mizda plateau, along the Nalut-Ghadames road, and at the northern (Homs) and southern end of the Hon graben (Jufrah oasis, Jabal Waddan, and Jabal as Soda).

Between Azizia and Mizda, along a north-south traverse, a denser spacing of sites was established. Between Garian and Jefren, the spacing of sta-

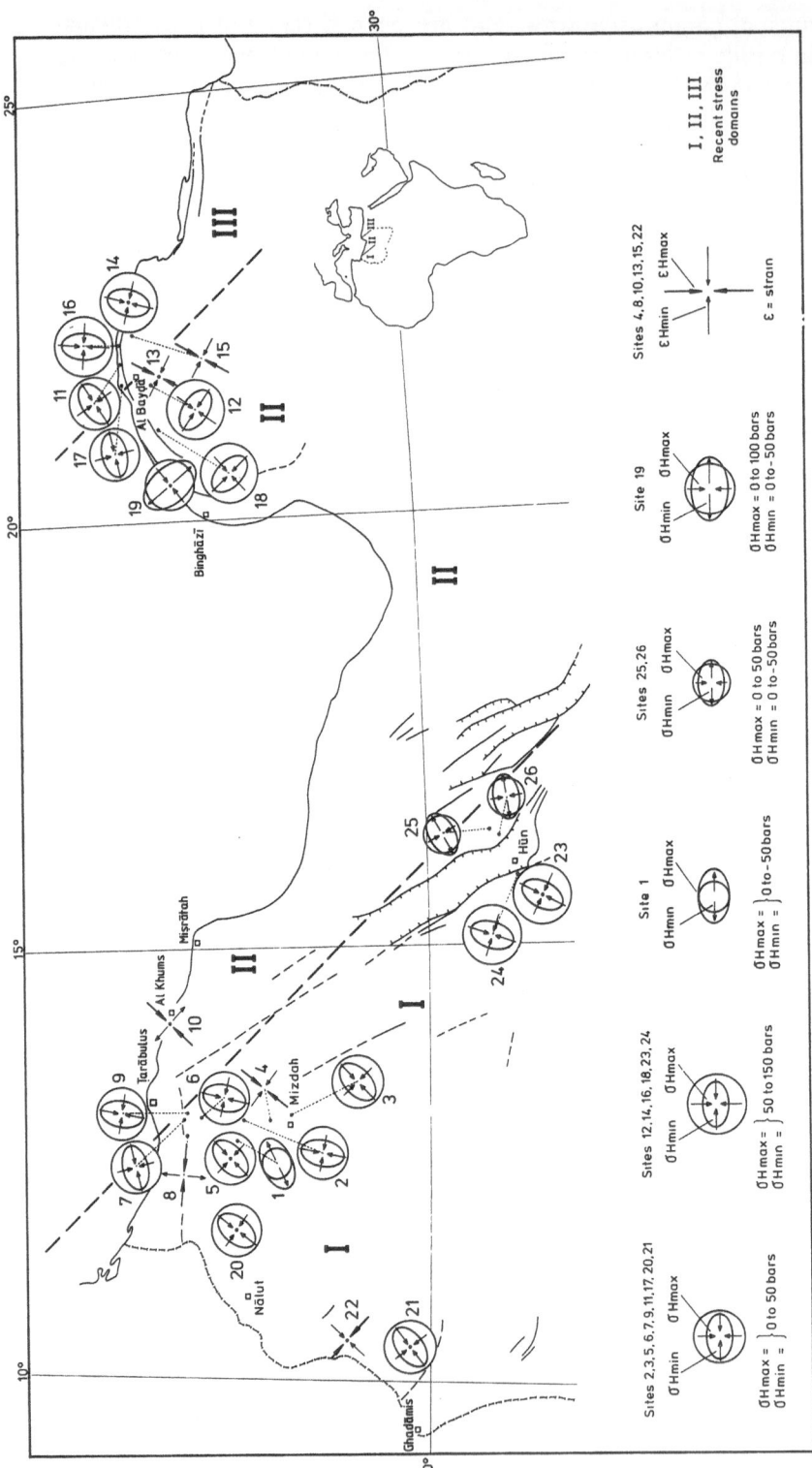

Fig. 1. Locations of *in situ* strain measurements in Libya. Stress ellipses show the orientations and amounts of actual horizontal principal stresses. The circles indicate a state of released strain after overcoring and surround the ellipses if σ_{Hmax} and σ_{Hmin} are compressive, while they cut through the ellipses if the present horizontal stress is tensile. The broken heavy lines separate Recent stress domains of consistent orientation of σ_{Hmax} and σ_{Hmin}. The young tectonics and the seismological events of Libya seem to be controlled by the actualrock stresses. Al Beda = Al Baydā, Beugari = Bīnghāzī, Ghadames = Ghadāmis, Homs = Al Khums, Hon = Hūn, Misurata = Misrātah, Mizda = Mizdah, Nalut = Nālut, Tripoli = Tarābulus

tions was also denser than usual (Fig. 1). The Cyrenaican sites were located between a point north of Bengasi and Derna and followed a line almost parallel to the coast. The north-south extension of sites in Cyrenaica was short since during this stage of *in situ* strain measurements in Libya it seemed more important to investigate the orientation of principal stress components in the east-west direction across Libya. The Carenaican sites are located in the coastal plain as well as on the plateau that follows the various escarpments within Al Jabal al Akhdar.

Temperature Induced Stresses

Since *in situ* strain measurements were performed at surfaces of horizontal rock exposure, daily and annual temperature variations that cause ground stress anomalies must be determined so as to calculate the pure tectonically induced stresses. If the period (P) and the amount (T_s) of the temperature variations at the surface and the diffusivity constant (K) are known, the depth of no temperature variation can be calculated by the equation (Carslaw and Jaeger, 1959):

$$T_x = T_s e^{-x\sqrt{\frac{\pi}{KP}}}$$

Hooker and Duvall (1971) have shown by field tests that the Mt. Airy granite (N. C., U. S. A.) and a granite at Marble Falls (Tex., U. S. A.) show daily temperature variations that were significant to a depth of about 0.50 m when the temperature variation of the air was 12^0 C. The annual temperature variations of those granites were significant to a depth of at least 8 m when the seasonal temperature variations (using the daily mean temperature at surface) was 27^0 C.

In Libya a greater annual air temperature variation (30^0 C) and a greater diffusivity constant for carbonates may increase the temperature penetration depth considerably. From July 21—23, 1977, a permanent field experiment was made at Socna (Jufrah oasis) to measure horizontal rock strain at the surface and corresponding rock temperatures. The strain difference between maximum expansion at the highest temperature (43^0 C, 2 p. m.) and the minimum expansion (or maximum contraction) at the lowest temperature (30^0 C, 7 a. m.) were about $110\ \varepsilon$. This strain variation under diurnal temperature changes was measured at 4 separate strain gauges bonded to the limestone at the surface. The temperature-induced strain variations were consistent in the 0^0 (ε_a)-, 90^0 (ε_c)-, and 135^0 (ε_b)-directions as shown by the strain gauges. Thus, the limestone at the Socna test site was isotropic in thermal expansion. No rotation of the tectonic strain ellipse must be expected nor any variation of the directions of principal strains.

To calculate the horizontal stress (σ_h) due to temperature changes ($T_1 - T_0$) the thermal coefficient of expansion (α), the elastic modulus (E), and the Poisson's ratio (v) of the rocks were determined in laboratory experiments. Thus, the results as shown in Table 1 were obtained using the formula of Timoshenko and Goodier (1951):

$$\sigma_h = \frac{\alpha\, E\, (T_1 - T_0)}{1 - \nu}$$

T_1 = temperature during *in situ* overcoring test.

T_0 = annual mean temperature at *in situ* test site.

Due to the α-values there was no thermally induced stress at the Socna test site. Since no data of field experiments were available, no thermal strain release at the other locations was considered. The temperature-induced

Table 1

Location (numbers refer to Fig. 1)	Thermal coefficient of expansion (α) 10^{-6} C^{-1}	Elastic modulus E 10^5 bar	Poisson's ratio (ν)	Temperature test site T_1	Annual mean temperature T_0	Location of annual mean temperature
1	10.0	5.2	0.25	26.5⁰ C	19.7⁰ C	Jefren
2		4.5	0.25	20.0⁰ C	20.0⁰ C	
3		3.5	0.25	21.5⁰ C	21.2⁰ C	
5		0.9	0.30	20.0⁰ C	19.2⁰ C	
6	6.0	4.0	0.25	25.0⁰ C	21.9⁰ C	Azizia
7		3.5	0.25	22.5⁰ C	21.9⁰ C	
9	6.0	5.2	0.25	29.5⁰ C	21.9⁰ C	Azizia
11	4.1	1.5	0.30	21.0⁰ C	16.3⁰ C	Cyrene
12	5.2	3.5	0.30	16.5⁰ C	17.9⁰ C	Marj
14	9.1	2.5	0.25	22.5⁰ C	20.8⁰ C	Derna
16	5.0	4.0	0.25	24.0⁰ C	20.8⁰ C	Derna
17	5.0	4.0	0.25	18.5⁰ C	16.3⁰ C	Cyrene
18	9.1	1.5	0.30	24.5⁰ C	17.9⁰ C	Marj
19	4.6	4.4	0.25	24.0⁰ C	20.5⁰ C	Bengasi
20	9.1	4.0	0.25	23.0⁰ C	20.0⁰ C	Nalut/Azizia
21	8.1	5.0	0.25	24.0⁰ C	22.1⁰ C	Ghadames
23	3.7	3.0	0.25	43.0⁰ C	21.1⁰ C	Hon
24	8.6	5.4	0.25	43 to 30⁰ C	21.1⁰ C	Hon
25	10.3	2.0	0.30	30.0⁰ C	21.1⁰ C	Hon
26	10.3	2.0	0.30	40.0⁰ C	21.1⁰ C	Hon

stresses (σ_h) with the exception of one site were all compressive, since T_1 exceeded or was equal to T_0. The tectonic stresses were obtained after substraction of thermally induced stresses from the total stress amounts.

Results and Interpretations

166 individual *in situ* strain measurements were performed after over-coring of carbonates at 25 sites in northern Libya. In addition, 211 readings of strain were made at one site (Socna, Jufrah oasis) during a field experiment to test temperature-induced stresses. Table 2 shows the computed maximum (ε_1) and minimum (ε_2) horizontal strains, the direction and amount of maximum horizontal stress ($\sigma_{H\max}$) and the minimum horizontal stress ($\sigma_{H\min}$). Also, the geographic description of the sites and the stratigraphic and lithological characteristics of the rocks have been added to Table 2.

In the Jefara plain west of Azizia and on the Jefren-Garian-Mizda plateau there is a remarkable consistency in the orientation and the amount of the principal horizontal stresses. $\sigma_{H\max}$ remains NW—SE-directed all over the central and western part of Tripolitania. Except at site 1, the tectonic stresses are all compressive and range between 0 and 50 bars (Fig. 1). NW—SE-directed horizontal compressive stress was also found at the southern end of the Hon graben. However, on the western side of the graben between Socna and the Jabal as Soda maximum horizontal stresses were oriented 110^0—120^0 and higher (50—160 bars) than on the eastern side of the Hon graben, where the $\sigma_{H\max}$-directions were 150^0—170^0 and the tectonic stresses 50 to −50 bars (Fig. 1). It is suggested that east of the Hon graben in the Jabal Waddan a change of the Recent stress regime is indicated by this decrease and by the clockwise rotation in the orientation of tectonic stresses. Towards the NW in eastern Tripolitania, there is evidence of this different stress domain (site 10, Fig. 1). The heavy broken line on Fig. 1 that runs NW−SE from Sabratah at the Tripolitanian coast to Azizia, Bu Ngem and the Jabal Waddan marks the area of transition between the western Recent stress domain of Libya (I) and a central Recent stress domain (II).

Cyrenaica was another area of *in situ* strain investigations (Fig. 1). The stresses there were generally higher than in Tripolitania. The western part of Cyrenaica is dominated by a Recent horizontal stress field whose $\sigma_{H\max}$ is NE−SW-oriented, suggesting that the entire Sirte basin and its Tripolitanian and Cyrenaican marginal areas possibly belong to the central Libyan *in situ* stress field (II, Fig. 1). Additional *in situ* strain measurements in the Sirte basin need to be carried out to prove if the central Libyan stress domain holds from Tripolitania to Cyrenaica. North and east of Al Beda (Cyrenaica) and mainly along the coastal area $\sigma_{H\max}$ again was NW−SE-oriented. It is suggested that eastern Cyrenaica is part of a Recent stress domain that probably extends much further to the east (III, Fig. 1). The shorter broken line in central Cyrenaica marks the transition between the central (II) and the eastern (III) Libyan stress domains.

Residual Stresses

Residual stresses may be locked into the rocks by the same variety of stress-producing processes as those that contribute to the existence of a Recent stress field. It is the residual stress of a palaeo-stress field that could

Table 2

Site	Location	Stratigraphy	Lithology	ε_1	ε_2	Direction of σ_{H}max	σ_{H}max bar	σ_{H}min bar
1	Quarry, 10 km E of Chicla	Ain Tobi Formation	Micrite 10°- and 110°-joint maxima	−2	−12	162°	−3	−7
2	21 km S intersection of road Mizda-Garian with road Garian-Jefren	Cenomanian	Reefal limestone dense with corals, 0°- and 80°—90°-joint maxima	77	46	105°	43	31
3	2,5 km E of Mizda	Upper Cretaceous	Porous limestone, 80°—90°- and 170°-joint maxima	123	82	141°	47	37
4	33 km N of Mizda	Upper Cretaceous	Siliceous, dolomitic limestone, pop-ups (80°-strike = 170°-compression)	621	189	42°		
5	7 km E of intersection roads Jefren-Garian-Giado	Ain Tobi Formation	Hard limestone at surface, soft below	301	34	139°	31	12
6	3 km N of granary at the road Azizia-Garian	Azizia Formation (Muschelkalk)	Hard micrite	69	54	107°	26	21
7	9 km S of Azizia, quarry 2 km E of the road Azizia-Garian	Azizia Formation (Muschelkalk)	Dolomitic micrite, 60°—70°-horizontal stylolites	43	−14	169°	13	−1
8	Quarry 9 km SW Azizia at the road to Jefren	Azizia Formation (Muschelkalk)	Dolomitic micrite	51	14	102°		
9	Quarry at Azizia	Azizia Formation (Muschelkalk)	Dolomitic micrite	76	55	103°	19	10
10	Quarry 10 km W Homs at the road to Tripoli	Upper Cretaceous	Dense limestone	122	−53	45°		
11	Road Apollonia-Cyrene, between 2. and 3. escarpment S of the coast (Cyrenaica)	Derna Formation (Eocene)	Breccia with micritic matrix	155	32	148°	23	9
12	10 km W of Salantah (Cyrenaica), road Salantah-Qandulah	Derna Formation	Solid nummulitic limestone	298	83	35°	128	70

No.	Location	Age / Formation	Lithology			°		
13	10 km S of Salantah (Cyrenaica) at Jardas al Jarari	Upper Cretaceous (? Eocene)	Solid limestone close to a fault	163	25	26°	61	41
14	3 km W of Derna (Cyrenaica)	Quaternary	Solid breccia	219	121	104°		
15	1 km W of Al Qubbah (Cyrenaica)	Oligocene	Marly limestone, measurement on hard caliche	340	144	29°		
16	9 km E of Apollonia, coastal road to Al Hilal	Derna Formation (Eocene)	Nummulitic limestone, 10°- and 98°-joint maxima	204	123	94°	92	66
17	22 km N of Al Beda (Cyrenaica), road Beda-Hamamah	Derna Formation (Eocene)	Solid nummulitic limestone, 100°-horizontal stylolites	80	15	173°	30	10
18	88 km W Al Beda, road Bengasi-Beda, close to Marj!	Quaternary	Breccia	875	91	50°	136	46
19	50 km E of Bengasi, road to Al Beda	Tukrah Formation (Senonian)	Solid limestone	280	-166	54°	101	-56
20	5 km E of Al Giosc	Quaternary	Solid conglomerate, micritic matrix	101	73	123°	37	28
21	5 km N of Derj	Upper Cretaceous	Solid dolomite, 70°- and 140°-joint maxima	104	60	144°	53	35
22	73 km N of Derj	Upper Cretaceous	Solid limestone, 40°—70°- and 140°- joint maxima, 10°- and 150°- horizontal stylolites	266	167	131°		
23	E margin of Jabal as Soda at the road to Sebha	Upper Cretaceous	Solid limestone, 70°—90°- and 140°- joint maxima	285	161	118°	72	42
24	1 km E of Socna, road to Hon, temperature test site	Upper Cretaceous	Solid micrite 30°- and 110°-joint maxima	410—311	323—263	110°	160	137
25	17 km E of Waddan (Jufrah oasis)	Eocene	Marly limestone	211	14	149°	21	-9
26	14 km E of Waddan	Eocene	Marly limestone	226	175	170°	6	-2

not be eliminated by the actual stress. Residual stresses can be recovered after a second overcoring of a rock specimen that has been released from the boundary loads of the present stress field by a previous overcoring.

For the first overcoring an 8 cm core barrel was used and for the second overcoring a 5 cm wide core bit was placed inside the 8 cm rock cylinder that was still in its original position (Engelder and Sbar, 1976).

Tests for recovering residual stresses were made in Tripolitania and Cyrenaica as well. At all sites an additional strain release after the second overcoring was observed. But only at site 3 (Fig. 1) were the directions of additional released strain (35^0) different from those of the present principal stress orientations (141^0). This single result of residual stress in Tripolitania supports the concept that a previous stress field with NE−SW-oriented $\sigma_{H\max}$ existed in Tripolitania during the Palaeogene. There is more evidence from the results of structural investigations in this area (Schäfer, 1980, this volume).

Tectonic and Seismological Implications

There are numerous structural features in western Tripolitania such as joints, veins, horizontal stylolites, volcanic dikes and faults that are NW−SE-oriented. They have been formed last among structures of other trends and tectonic regimes in that area. The presently active tectonic crustal stresses in western Tripolitania could have formed these NW−SE-striking features. This would imply that the Recent stress domain was already active during the Pliocene since the volcanic dikes were intruded during that period. There is evidence in eastern Tripolitania that the youngest structures are NE−SW-directed. The age of their formation is not known but they must have been created by a stress field that is consistent with the present one.

The Hon graben appears morphologically significant only between the Jufrah oasis in the south and Bu Ngem in the north. The prolongation of the graben's bordering master faults from there further to the north are shown on the geological map of Libya (Conant and Goudarzi, 1964) as far north as the Jefara plain. But the Upper Cretaceous strata between Tarhuna and Homs do reveal only two normal faults that have throws of a few metres. I suggest that young subsidence of the Hon graben beyond the limits of the present stress field of western Libya (I) was not possible because of the NE−SW-oriented $\sigma_{H\max}$ of the central Libyan stress field (II).

In Cyrenaica the two main trends of maximum horizontal stress separating the central from the eastern Libyan stress domain may have an impact on the young activity of many major faults. These faults are considered to have already been normal faults in the Palaeogene or earlier. Their strike is about parallel to the coast line and is consistent with the orientation of $\sigma_{H\max}$ in western and eastern Cyrenaica.

The high *in situ* stress in Cyrenaica compared to other areas of northern Libya may also be responsible for the increased seismic activity. The last major event was the earthquake of 1963 that destroyed the town of Marj. The epicentre was located NE of Marj and close to one of these major faults that form northfacing escarpments between the Jabal al Akhdar and the coast. A solution of the focal mechanism of that earthquake is not available but according to the measured magnitude of 5.3 (Campbell, 1968) normal faulting alone may not be sufficient to explain the amount of released focal energy.

Also close to the transition zone between two stress domains is a NW−SE-oriented cluster of epicentres about 100 km south of Misurata at Al Ghadenia. In 1977, Tripoli was struck by an earthquake of minor magnitude. It is likely that there is a stress concentration along the transition zones (broken heavy lines on Fig. 1) causing a higher risk of earthquakes than elsewhere in Libya. *In situ* stress site 18 which was the nearest of all sites to Marj showed the highest absolute compressive stress in Cyrenaica.

Plate Tectonic Implications

Since the *in situ* rock stresses of major parts of northern Libya are of tectonic origin as previously concluded, we have to find a plate tectonic explanation for their formation and orientation. It is suggested that the *in situ* stresses mainly derive from plate movement and plate collision. According to Burke and Wilson (1972) the African plate is stationary since 30 m. y. or has had a slight northward movement since 10 m. y. (Pitman and Talwani, 1972). Also, Europe is stable north of the Pyrenees, Alps and Carpathian Mountains. Thus, it is difficult to present a plausible conclusion for the considerable amount of NW−SE crustal shortening that occurred only during the Neogene and is still effective in the Mediterranean orogenic belts. Structural studies show an African-European NW−SE-approach of many hundreds of kilometres during the Neogene (Schäfer, 1980, this volume) also causing the western and eastern Libyan stress domains with $\sigma_{H\max}$ to be oriented in the same NW−SE-direction.

It is suggested that Tripolitania and eastern Cyrenaica have northward-extended crustal spurs that accumulate NW−SE-directed $\sigma_{H\max}$ during the collision while the areas at both sides of the rigid spurs and their extensions directed further inland are under minor stress. The central Libyan stress domain may be considered as a secondary stress field resulting from a scissors-like closure due to the Tripolitania-Cyrenaica approach.

Conclusions

In situ strain measurements performed in northwestern and southern Tripolitania show a systematic arrangement of Recent horizontal principal stress orientations. The maximum horizontal stress ($\sigma_{H\max}$) is NW−SE-directed and is always compressive, while normal to this direction the minimum horizontal stress ($\sigma_{H\min}$) is mostly compressive, too. At only a few sites is $\sigma_{H\min}$ tensile. This area has been named the western Libyan stress domain. In a study of the seismicity and earthquake mechanisms of the Mediterranean Ritsema (1975) suggested that all Maghreb countries except the western part of Morocco are subjected to a Recent NW−SE-oriented horizontal compressive stress. The western Libyan stress domain is, thus, the eastern extension of a large *in situ* stress regime.

Eastern Tripolitania, the Sirte basin and western Cyrenaica are subjected to Recent horizontal stresses that show a different orientation from those of the western Libyan stress domain. There, the maximum horizontal stress is NE−SW-directed and is compressive, and, thus, will presently prevent the northern Hon graben and the Sirte grabens from active subsidence.

The central Libyan stress domain ends to the east along a zone that can be traced from Al Beda to Al Makili in Cyrenaica and is followed by the eastern Libyan stress domain with NW−SE-directed compressive $\sigma_{H\max}$.

The seismic activity in Libya is concentrated on the central Hon graben and Cyrenaica, and the locations of epicentres match fairly well the transition zones of Recent stress domains where stress accumulation may be assumed.

From the knowledge of the present stress distribution in surface rocks it is possible to focus on the Recent stress situation in subsurface rocks. The actual stress conditions in source and reservoir rocks of fluids can be deduced from surface stress results. Migration and accumulation of fluids in rocks are considerably controlled by the present state of rock stress that modifies porosity and permeability.

Acknowledgements

This study was financed by the National Oil Corporation of the S. P. L. A. J. (Libya). Marengo Ltd. under the guidance of W. H. Kanes, Columbia, S. C., U. S. A., supported the project with administrative help and constructive discussions. J. Erdmann, H. Häusler, and K.-H. Kraft, all of Karlsruhe, Germany, kindly helped to perform the fieldwork. I have to thank K. Balthasar, Karlsruhe, for his aid in determinations of the elastic moduli and Poisson's ratios.

References

Burke, K., Wilson, J. T.: Is the African Plate Stationary? — Nature *239*, 387—390 (1972).

Campbell, A. S.: The Barce (Al Marj) Earthquake of 1963. In: F. T. Barr (ed.), Geology and Archaeology of Northern Cyrenaica, Libya. Petrol. Explor. Soc. Libya, 10. annual field conf., 183—195 (Holland-Breumelhof N. V.), Amsterdam (1968).

Carslaw, H. S., Jaeger, J. C.: Conduction of Heat in Solids. 2. ed., 510 pp. (Clarendon Press), London (1959).

Conant, L. C., Goudarzi, G. H.: Geological Map of Libya, 1 : 2 Mill., U. S. G. S., Misc. Geol. Invest. Map 1—350 A, Wash. D. C. (1964).

Engelder, J. T., Sbar, M. L.: Evidence for Uniform Strain Orientation in the Potsdam Sandstone, Northern New York, From In Situ Measurements. J. Geophys. Res. *81*, 3013—3017 (1976).

Haimson, B., Fairhurst, C.: In-situ Stress Determination at Great Depth by Means of Hydraulic Fracturing in Rock Mechanics — Theory and Practice. Proc. 11. Symp. Rock Mechanics (ed. W. H. Somerton), Am. Inst. Mining Engineers, 559—584 (1970).

Hooker, V. E., Duvall, W. I.: In Situ Rock Temperature — Stress Investigations in Rock Quarries. Bu-Mines Rept. of Inv. 7589, 12 pp. (1971).

Leeman, E. R.: The "Doorstopper" and Triaxial Rock Stress Measuriug Instrument Developed by the C. S. I. R. — J. S. Afr. Min. Metall. 69, 305—339 (1969).

Pitman, W. C., Talwani, M.: Sea-Floor Spreading in the North Atlantic. Bull. Geol. Soc. Am. 83, 619—646 (1972).

Ritsema, A. R.: The Contribution of the Study of Seismicity and Earthquake Mechanisms to the Knowledge of Mediterranean Geodynamic Processes. In: Progress in Geodynamics, Geodyn. Proj., Sci. Rep. 13, Borradaile et al. (eds.), 142—153 (North Holland Pub. Co.), Amsterdam (1975).

Schäfer, K.: Geodynamik an Europas Plattengrenzen. Fridericiana 23, 30—46 (1978).

Schäfer, K.: Paleo- and Recent Stress Fields in Libya and Tunisia from the Cenozoic Structural Bearing. This volume (1980).

Timoshenko, S., Goodier, J. M.: Theory of Elasticity, 2. ed., 506 pp., (McGraw-Hill), New York (1951).

Turcotte, D. L., Oxburgh, E. R.: Stress Accumulation in the Lithosphere. Tectonophysics 35, 183—199 (1976).

Address of author: Prof. Dr. Karlheinz Schäfer, Institut für Geowissenschaften der Universität Bayreuth, D-8580 Bayreuth, Federal Republic of Germany.

Rock Mechanics, Suppl. 9, 63—68 (1980)

Rock Mechanics
Felsmechanik
Mécanique des Roches
© by Springer-Verlag 1980

Theme 2

Stresses Inferred from Fault Plane Solutions of Earthquakes

Crustal Stresses Inferred from Fault-Plane Solutions of Earthquakes and Neotectonic Deformation in Switzerland*

By

N. Pavoni

With 2 Figures

Abstract

P-axes inferred from fault-plane solutions of 23 recent earthquakes in Switzerland and adjacent areas reveal a NNW – SSE (E Switzerland) to NW – SE and WNW – ESE (W Switzerland) orientation of maximum horizontal compressive stress in the upper crust. Nearly the same orientation of maximum compression is indicated by in situ stress measurements. The orientation of maximum horizontal compressive stress corresponds well with the orientation of maximum horizontal crustal shortening as derived from a kinematic analysis of Neogene and Quaternary structural features. Evidently the stress field which causes the present seismicity is very similar in its orientation to the stress field of the last 5 to 10 million years which produced the neotectonic deformation. In several cases of earthquakes in the Helvetic Alps, the Lake of Constance area and in the Jura Mountains a distinct relation between the parameters of the fault-plane solution and neotectonic deformation is shown up.

1. Introduction

The focal mechanisms of 23 earthquakes in Switzerland and adjacent areas were investigated systematically. The earthquakes occurred in the years 1961 to 1979. Their magnitudes M_L range from 2.5 to 5.2. During the last several years the station network of the Swiss Seismological Service was enlarged considerably. The increased number and regular distribution of seismic stations in Switzerland allow in many cases the focal mechanism of even low magnitude earthquakes to be evaluated. In addition, several micro-earthquake surveys with special regard to the relationship of seismicity to local tectonic deformation have been performed in seismically active areas, such as the central Valais area.

---

* Contribution No. 261 of Institut für Geophysik, ETH Zurich.

0080-3375/80/Suppl. 9/0063/$ 01.20

Table 1. List

No.	Date (GMT)	h m s	Lat./Long.	Mag.
1	April 28, 1961	20 : 48 : 49	47.7 N/7.9 E	4.9
2	March 14, 1964	02 : 37 : 22	46.9 N/8.3 E	5.2
3	February 5, 1968	02 : 28 : 49	46.6 N/5.8 E	3.5
4	June 21, 1971	07 : 25 : 29	46.4 N/5.8 E	4.4
5	September 29, 1971	07 : 18 : 52	47.1 N/9.0 E	4.8
6	May 8, 1973	19 : 08 : 24	45.6 N/9.7 E	3.9
7	July 9, 1973	00 : 27 : 06	46.8 N/9.7 E	3.8
8	January 19, 1974	02 : 49 : 52	46.7 N/7.5 E	3.8
9	April 26, 1974	07 : 21 : 07	47.2 N/7.9 E	3.0
10	May 21, 1975	04 : 10 : 47	45.8 N/8.9 E	2.5
11	December 29, 1975	05 : 25 : 17	47.1 N/9.2 E	3.1
12	January 29, 1976	11 : 39 : 08	46.3 N/7.5 E	3.6
13	March 2, 1976	08 : 27 : 57	47.6 N/9.4 E	3.7
14	March 22, 1976	14 : 44 : 23	47.0 N/7.0 E	2.7
15	March 26, 1976	22 : 28 : 31	47.6 N/9.5 E	4.1
16	June 9, 1974	00 : 18 : 09	46.0 N/6.4 E	3.6
17	May 29, 1975	00 : 32 : 39	46.0 N/6.0 E	4.2
18	July 27, 1976	17 : 51 : 56	45.9 N/6.7 E	3.1
19	November 21, 1977	19 : 27 : 42	47.3 N/8.5 E	3.5
20	August 13, 1978	04 : 02 : 27	47.3 N/7.8 E	3.4
21	August 28, 1978	16 : 44 : 41	47.3 N/8.2 E	2.8
22	December 4, 1978	23 : 54 : 24	46.6 N/6.9 E	3.3
23	July 3, 1979	21 : 13 : 12	47.0 N/7.2 E	3.8

2. Fault-Plane Solutions and Neotectonic Deformation

The earthquakes as well as the P- and T-axes derived from the fault-plane solutions are listed in Table 1. Their location and type of fault-plane solution as well as the orientation of the P-axis are shown in Fig. 1. The foci range from 2—20 km in depth. The P-axes arrange themselves in a regular pattern as indicated by the dashed lines in Fig. 1. Basically, a fault-plane solution represents a movement picture, the P-axis corresponding to the axis of maximum shortening in the focus. Under the assumption that the axis of maximum shortening corresponds to the axis of maximum compressive stress, the P-axes in Fig. 1 reveal a regular stress field in the upper crust of the central and northern Swiss Alps, the Swiss Plateau, and the Jura Mountains.

The orientation of maximum horizontal compressive stress, inferred from the orientation of P-axes, is NNW − SSE in eastern Switzerland, to NW − SE and WNW − ESE in western Switzerland. It corresponds well with the orientation of maximum horizontal crustal shortening derived from a

Fig. 1. Fault-plane solutions of 23 earthquakes in Switzerland and adjacent areas, listed in Table 1. Equal area lower hemisphere projections. Dark quadrant indicates compression, light quadrant dilatation. Arrows: Orientation of P-axis. Dashed lines indicate orientation of maximum horizontal compression in the upper crust. M: Composite solution from microearthquake survey in 1976

of Earthquakes

Region	P	T	References
Schopfheim	134/18	38/18	Ahorner and Schneider, 1974
Sarnen	139/11	37/44	Ahorner et al., 1972
Clairvaux	134/7	314/83	Pavoni and Peterschmitt, 1974
Jeurre	315/32	54/14	Pavoni and Peterschmitt, 1974
Glarus	158/0	68/6	Mayer-Rosa and Pavoni, 1977
Bergamo	297/22	43/35	Pavoni et al., 1977
Arosa	156/3	66/3	Mayer-Rosa and Pavoni, 1977
Berner Alpen	124/5	214/5	Mayer-Rosa and Pavoni, 1977
Berner Mittelland	354/18	240/51	Mayer-Rosa and Pavoni, 1977
Varese	351/4	82/19	Pavoni et al., 1977
Glarus	311/14	41/1	Pavoni et al., 1977
Valais	111/58	8/10	Mayer-Rosa and Pavoni, 1977
Bodensee I	346/0	76/0	Mayer-Rosa and Pavoni, 1977
St. Blaise	148/0	58/0	Pavoni, 1977 b
Bodensee II	348/0	78/0	
Faucigny	96/0	6/0	Fréchet, 1978
Vuache	102/15	193/6	Fréchet, 1978
Faucigny	85/0	175/0	Fréchet, 1978
Albis	169/0	79/0	
Balsthal	156/0	66/0	
Hinwil	166/0	76/0	
Semsales	135/0	45/0	
Murten	150/0	60/0	

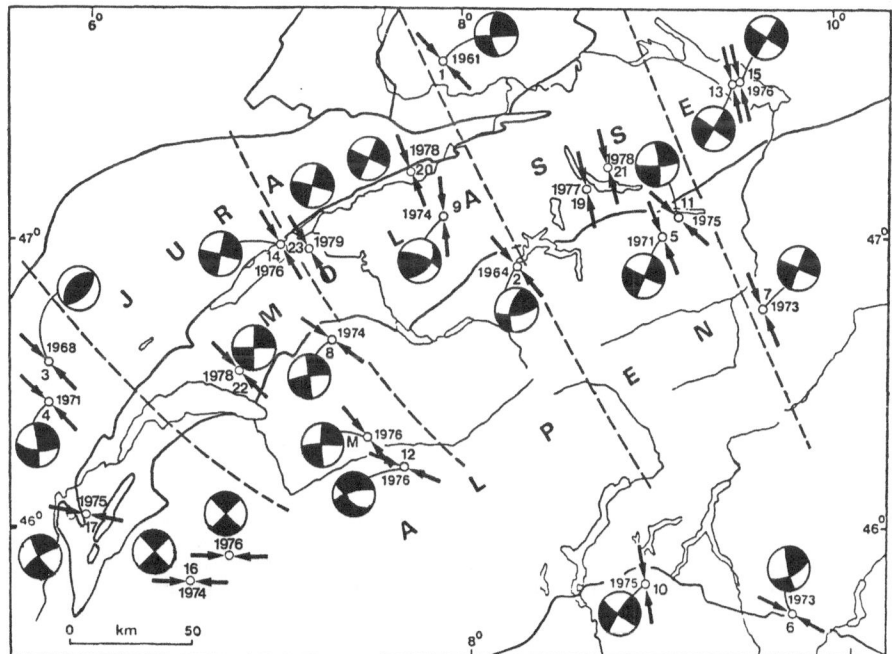

Fig. 1

kinematic analysis of neotectonic, i. e. Neogene and Quaternary, structural features as shown on the seismotectonic map of Switzerland (Pavoni and Mayer-Rosa, 1978). The observed rotation of the orientation of maximum compressive stress follows the rotation of the strike of the Alpine arc. The data from in situ stress measurements (Illies and Greiner, 1978; Kovari et al., 1972; Gysel, 1975) are consistent with the seismological findings. The new seismological data confirm the conclusion that, with regard to the orientation of maximum horizontal compressive stress, the stress field which causes the seismicity in Switzerland is very similar to the stress field of the last 5 to 10 million years which produced the neotectonic deformation.

3. Seismicity and Local Fault Systems

It is very difficult and in most cases impossible to attribute a single seismic event with certainty to a distinct known fault, unless the fault movement reaches the earth's surface. The reasons for these difficulties are evident. In a few kilometers depth the conventional tectonic picture as derived from surface geology is often considerably modified. There are good reasons to

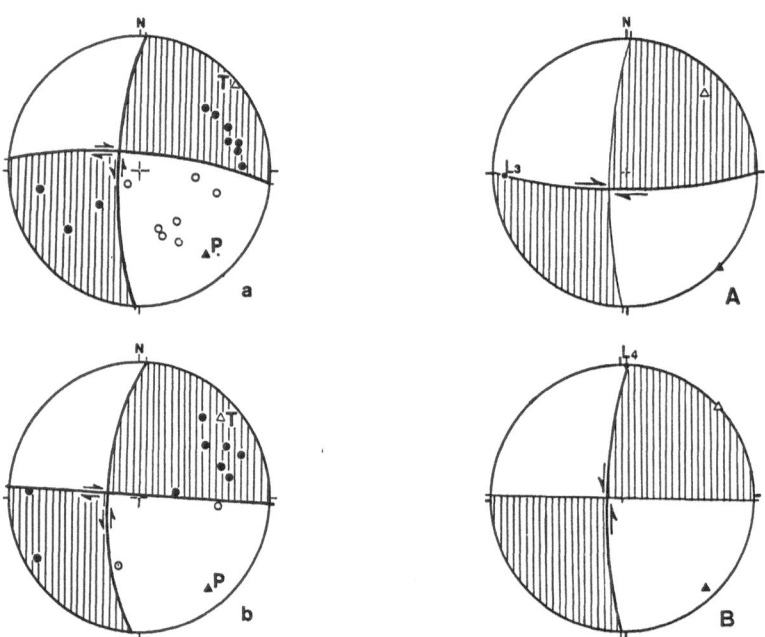

Fig. 2. Two composite fault-plane solutions (a, b) from microearthquakes of the Wild-hornzone, May—June 1976, after Pavoni (1977 a) and neotectonic deformation (A, B) of the western Aarmassif after Steck (1968), in fault-plane solution representation. Equal area lower hemisphere projection. Hatched: Compressional quadrant; light: Dilatational quadrant. The nodal planes indicate N-S sinistral and E-W dextral strike-slip movements. P: P-axis, T: T-axis, L_3: Lineation on E-W striking, steeply south dipping, fault-plane with dextral strike-slip displacement. L_4: Lineation on N-S striking steeply dipping fault-plane with sinistral strike-slip displacement. Solid triangle: Axis of maximum shortening. Open triangle: Axis of maximum lengthening

assume that there is a more intensive, small scale fracturing of the rocks than shown on geological maps. The accuracy of location of earthquakes is insufficient.

However, in several cases of earthquakes in the Helvetic Alps, the Lake of Constance area and in the Jura Mountains, there are clear indications from focal mechanism studies and epicenter locations that seismic acitivity is associated with certain fault zones or fault systems. In order to directly compare the geological and seismological data a method was developed to translate the geological observations into the seismological language (Pavoni, 1980). Fig. 2 shows two composite fault-plane solutions of microearthquakes of the Wildhornzone in the central Valais (Pavoni, 1977a) together with a representation of late-Alpine deformation structures observed in the western Aarmassif by Steck (1968), reproduced in fault-plane solution "language". The striking similarity of the results is an indication that the present tectonic movements, indicated by the seismic activity of the Wildhornzone, occur along the same systems of faults as the young fault systems found in the western Aarmassif.

References

Ahorner, L., Murawski, H., Schneider, G.: Seismotektonische Traverse von der Nordsee bis zum Apennin. Geol. Rdsch. *61, 915*—942 (1972).

Ahorner, L., Schneider, G.: Herdmechanismen von Erdbeben im Oberrhein-Graben und in seinen Randgebieten. In: Illies, H., Fuchs, K. (eds.): Approaches to Taphrogenesis (p. 104—117), Schweizerbart, Stuttgart (1974).

Fréchet, J.: La sismicité du Sud-Est de la France, et une nouvelle méthode de zonage sismique. Thèse 3e cycle, Université de Grenoble (1978).

Gysel, M.: In-situ Stress Measurements of the Primary Stress State in the Sonnenberg Tunnel in Lucerne, Switzerland. Tectonophysics *29*, 301—314 (1975).

Illies, J. H., Greiner, G.: Rhinegraben and the Alpine System. Geol. Soc. Am. Bull. *89*, 770—782 (1978).

Kovari, K., Amstad, Ch., Grob, H.: Ein Beitrag zum Problem der Spannungsmessung im Fels. Proc. Internat. Symp. f. Untertagebau, Lucerne, 501—512 (1972).

Mayer-Rosa, D., Pavoni, N.: Fault Plane Solutions of Earthquakes (1971—1976) in Switzerland. Proc. 15th E. S. C. gen. Assem. Krakow 1976, Publ. Inst. Geophys. Pol. Acad. Sc., *A-5 (116)*, 321—326 (1977).

Pavoni, N.: An Investigation of Microearthquake Activity in the Central Valais (Swiss Alps). Proc. 15th E. S. C. gen. Assem. Krakow 1976. Publ. Inst. Geophys. Pol. Acad. Sc., *A-5 (116)*, 317—320 (1977a).

Pavoni, N.: Erdbeben im Gebiet der Schweiz. Eclogae geol. Helv., *70/2*, 351—370 (1977b).

Pavoni, N.: Focal Mechanisms of Earthquakes and Neotectonic Deformation in the Helvetic Zone and the Aarmassif (Swiss Alps). Eclogae geol. Helv., *73/2*, (in prep., 1980).

Pavoni, N., Losito, G., Mayer-Rosa, D.: A Study of Focal Mechanisms of 1971—1976 Earthquakes in Switzerland and Northern Italy. Unpubl. report (1977).

Pavoni, N., Mayer-Rosa, D.: Seismotektonische Karte der Schweiz 1 : 750000. Eclogae geol. Helv. *71/2,* 293—295 (1978).

Pavoni, N., Peterschmitt, E.: Das Erdbeben von Jeurre vom 21. Juni 1971 und seine Beziehungen zur Tektonik des Faltenjura. In: Illies, J. H., and Fuchs, K. (eds.): Approaches to Taphrogenesis (p. 322—329), Schweizerbart, Stuttgart (1974).

Steck, A.: Die alpidischen Strukturen in den Zentralen Aaregraniten des westlichen Aarmassivs. Eclogae geol. Helv. *61/1,* 19—48 (1968).

Address of author: N. Pavoni, Institut für Geophysik der ETH, ETH-Hönggerberg, CH-8093 Zürich, Switzerland.

Rock Mechanics, Suppl. 9, 69—73 (1980)

Rock Mechanics
Felsmechanik
Mécanique des Roches
© by Springer-Verlag 1980

Seismic Stresses in Southern Germany

By

Götz Schneider

With 4 Figures

Abstract

Southern Germany as a tectonic and seismotectonic unit has the shape of a triangle. The borders of this area are formed by the eastern main fault of the Upper Rhinegraben, the chains of the folded Molasse layers in the north of the alpine nap area and the fractures at the southwestern edge of the Bohemian massiv. The most epicenters of this region are alined in zones striking NNE ("Rhenish direction") or NW ("Hercynian direction"). As can be deduced from fault-plane solutions, the prominent type of seismotectonic motions consists of horizontal strike slip motions along those directions. Measurements made with a larger number of mobile and permanent seismological stations in the most active area of this region, the western part of the Swabian Jura, confirm the a/m results. The observations made to date indicate that the direction of P_1 (P_1 = largest principal stress) is about NNW. One can see a tendency for a more northern direction of P_1 if one goes in Southern Germany from west to east. A comparison will be made between tectonic and seismotectonic observations.

1. Introduction

The most important earthquake sources in Southern Germany are two epicentral areas in the region of the Swabian Jura. The latter forms a part of the Southern German Triangle. This tectonic unit is bordered by the Upper Rhinegraben in the West, by the folded Molasse chains in the South and by some important fracture zones in the Northeast separating the triangle from other tectonic units in Thuringia and Bohemia (Fig. 1). Comparative tectonic studies could show that the most important structural element of the Southern German Triangle consists of widely spaced up- and down-warping of the crustal surface layers. These deformations show a clear relationship between the Southern German Triangle and the neighboring structures such as the Rhinegraben and the Alps. The southern part of the triangle is formed by the Molasse basin. The down-buckling of this foreland trough can still be recognized in the central part of the triangle. This tendency of deformation is superimposed upon the upwarping of the crystalline shield of the Black Forest and the Odenwald, forming a kind of compensation for the downgoing slab of the upper Rhinegraben.

0080-3375/80/Suppl. 9/0069/$ 01.00

Besides these deformations of large wave length, a system of synclines and anticlines showing wave lengths of about 30 km characterizes the status of deformation in the central part of the Southern German Triangle. This picture of more or less widely spaced buckling is completed by a system of

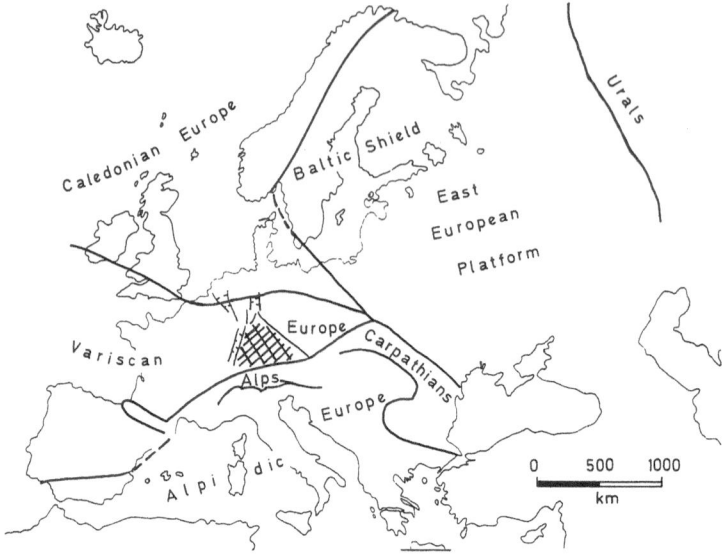

Fig. 1. Tectonic units in Europe

normal faulting forming especially some NW—SE striking graben zones in the middle part of the studied area (Carlé, 1955).

2. Seismicity and Present-Day Stress Field

Two directions of preferred epicenter alignment can be found in Southern Germany: The "Hercynian" (NW−SE) and the "Rhenish" (NNE) direction (Fig. 2). Considering these two strike angles as an expression for complementary shear fractures in the deeper crust, the main principal stress direction should follow the half angle between the two mentioned strike angles.

3. Fault-Plane Solutions

A large number of fault-plane solutions has been determined especially for the areas with high activity, such as the western part of the Swabian Jura, Upper Souabe and the Black Forest (Hiller, 1936 a and b; Schneider et al., 1966; Schneider, 1971; Haessler et al., 1979).

By far the most solutions show the character of horizontal strike-slip motions. Especially in the western part of the Swabian Jura, the evaluation of different fault-parameters together with fault-plane solutions indicates that the motions in the upper crust of this area belong to the type of left-

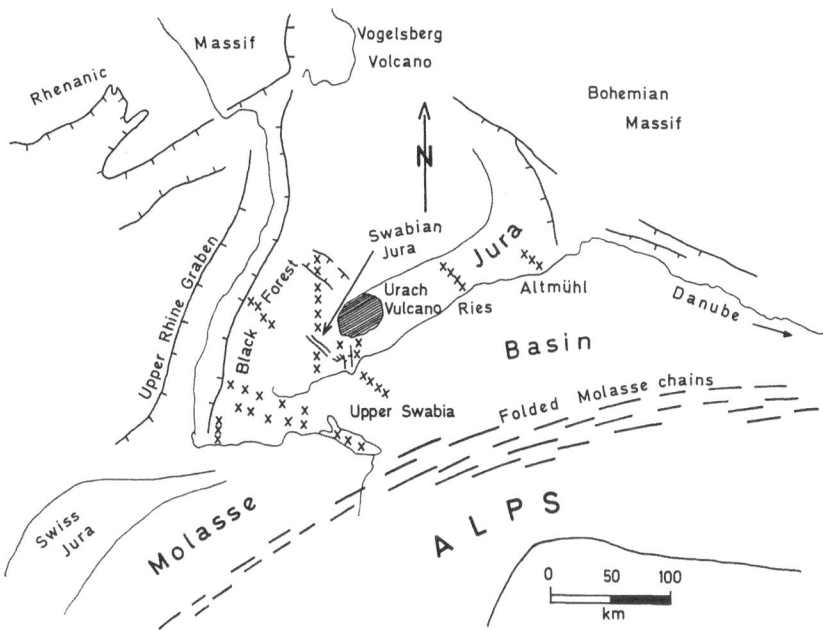

Fig. 2. "Hercynian" and "Rhenish" seismotectonic directions (crosses)

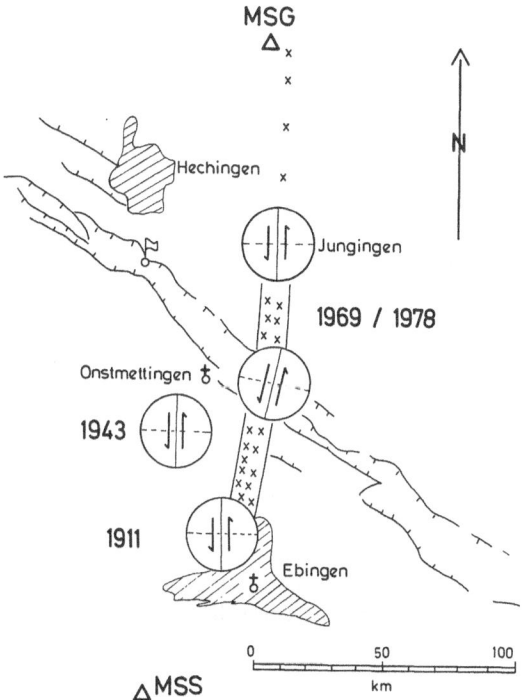

Fig. 3. Seismic block motions in the Swabian Jura

handed horizontal block motions striking N...NNE. This kind of motion can be detected over a distance of about 30 kilometers (Fig. 3). For all other epicentral areas, the information consists of fault-plane solutions and hypocentral coordinates only and is therefore too scanty to be interpreted directly in terms of tectonic motions in the crust.

4. Stress Field

From the distribution of epicenters and the orientation of fault planes it can be concluded that the axis of main principal stress is oriented about NNW. The existing information indicates that the direction of largest prin-

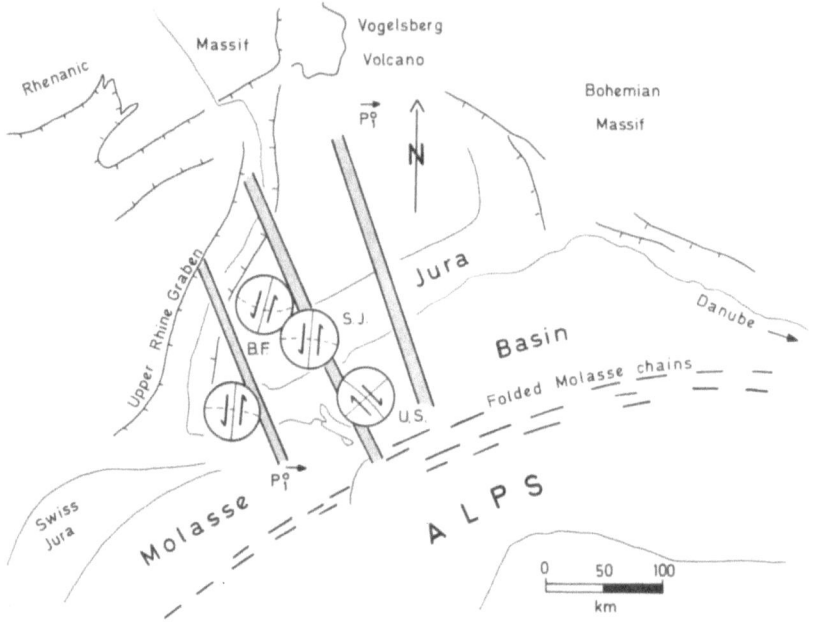

Fig. 4. Directions of largest principal stresses in S. Germany

cipal stress (P_1 in Fig. 4) changes its orientation slightly in the same sense as the normal to the general direction of the alpine chains.

5. Tectonic and Seismotectonic Notions

A comparison between tectonic and seismotectonic information leads to the conclusion that the recent seismotectonic block movement, concentrated to a depth interval between the crystalline basis and about 15...20 km, is the typical reaction of the upper-crustal layers to the general tectonic stress field.

The tectonic surface features as up- and down-warping of the sedimentary layers, the normal faults and graben formations are a secondary reaction on the block movements in the deeper crust. It could be demonstrated for other tectonic areas by Pavoni (1967) and Koide and Bhattacharji (1977) that such relations between tectonic motions in different floors of the crust exist.

References

Carlé, W.: Bau und Entwicklung der südwestdeutschen Großscholle. Beih. Geol. Jahrb. *16*, 272 pp. (1955).

Haessler, H., Hoang-Trong, P., Schick, R., Schneider, G., Strobach, K.: The September 3, 1978 Swabian Jura Earthquake (1980).

Hiller, W.: Das Oberschwäbische Erdbeben am 27. Juni 1935. Württbg. Jb. f. Statistik und Landeskunde, Jahrg. *1934/35*, 209—226 (1936a).

Hiller, W.: Das Hornisgrinde-Beben. Seism. Ber. d. Württbg. Erdbebenwarten, *1935*, Anh. 10—23; and Geol. Rundschau 27, 207—209 (1936b).

Koide, H., Bhattacharji, S.: Geometric Patterns of Active Strike Slip Faults and their Significance as Indicators for Areas of Energy Release. Energetics of Geological Processes (Editors: S. K. Saxena and S. Bhattacharji). Springer-Verlag, Berlin — Heidelberg — New York. p. 46—66 (1977).

Pavoni, N.: Kriterien zur Beurteilung der Rolle des Sockels bei der Faltung des Faltenjuras. Étages tecton. (Neuchâtel, 1967), 307—314 (1967).

Schneider, G., Schick, R., Berckhemer, H.: Fault-plane Solutions of Earthquakes in Baden-Württemberg. Zeitschr. f. Geophys. *32*, 383—393 (1967).

Schneider, G.: Seismizität und Seismotektonik der Schwäbischen Alb. Ferd. Enke Verlag, Stuttgart. 79 pp. (1971).

Address of author: Prof. Dr. Götz Schneider, Institute of Geophysics, University of Stuttgart, D-7000 Stuttgart, Federal Republic of Germany.

Rock Mechanics, Suppl. 9, 75—84 (1980)

Rock Mechanics
Felsmechanik
Mécanique des Roches
© by Springer-Verlag 1980

Seismic Stresses
in the Region Azores — Spain — Western Mediterranean

By

Agustín Udías

With 4 Figures

Abstract

The western end of the contact between the Eurasian and African plates extends to the Azores triple junction. Seismically, therefore, the region from the Azores islands to Italy must be considered together. At the western end the stresses are tensional normal to the axis of the ridge. Along the Azores-Gibraltar fault the motion is strike-slip right lateral changing to a north-south compression at the gulf of Cadiz. East of the strait, seismic activity is spread through the Alboran basin, Betica and Riff ranges. The continuation along the north African coast links this activity with the Sicily-Calabrian arc and Italian earthquakes. Stresses are mostly of compressional character, horizontal and normal to the trend of the activity. Main tectonic motions are the result of the main relative movements of the Eurasian and African plates, either compressional or right lateral.

Introduction

The southwestern part of the Eurasian plate and its contact margins with the American and African plates are zones of great interest from the point of view of geotectonics. We have limited our study to the region defined by latitudes 30^0 N to 50^0 N and longitudes 20^0 E to 40^0 W. This region includes a great variety of tectonic features. The western margin of the Eurasian plate is formed by a section of the mid-Atlantic ridge north of the Azores islands, trending roughly N−S. At the Azores islands there is a triple junction with ridge structure in its three branches. The eastern branch is formed by the Terceira ridge and ends abruptly at 24^0 W. From this point to the strait of Gibraltar, runs a W−E trending transcurrent fault. East of Gibraltar the rather simple contact of oceanic structures becomes a very complicated system of continental blocks, basins and orogenic belts. The westernmost continental block is the Iberian peninsula, with the two Alpine orogenies of the Betica and Pyrenees as its southeast and northeast limits. Between the continental parts of Africa and Europe lies a series of basins, from west to east, Alboran, Balearic, separated into two parts

0080-3375/80/Suppl. 9/0075/$ 02.00

by the Balearic islands, Ligurian and Tyrhenian. North of the Italian peninsula is located the orogenic belt of the Alps. Along the Italian peninsula runs the chain of the Apennines, which ends to the south in the Sicily-Calabrian arc, a subduction zone. The northern margin of the African plate is accidented by the Riff mountains which forms the southern part of the Gibraltar arc, extending east through the northern coast of Algeria and Tunisia by the Tell Atlas, to end also and the Calabrian arc.

The area has been the subject matter of numerous studies of geo-tectonic nature, the more recent interpretating the dynamics in the light of plate tectonics theory. Among those more closely related with the seismo-tectonic aspects of the region are Isacks et al. (1961), Ritsema (1969), McKenzie (1972), Payo (1972), Papazachos (1973), Udías et al. (1976), Hatzfeld (1978). In this paper the distribution of stresses and the relative motions of the plates in this region will be studied from seismic evidence, with particular emphasis on the boundary between the African and Eurasian plates from the Azores junction to the Calabrian arc.

Seismicity of the Region

The distribution of epicenters of this region for the period 1910—1977 as given from the Earthquake Data File of the NGSC, EDS, NOAA, is shown in Fig. 1. At the western part, epicenters outline the inverted Y shape of the mid-Atlantic ridge near the Azores Islands. The seismic zone, corresponding to the Azores-Gibraltar fault, extends from the end of the Terceira ridge to near 12^0 W. On this line large earthquakes occur $(M > 7)$, five in the 1910—1977 period, but the fault lacks a continuous low level activity. To the south, there is indication of a secondary south-east trending line, passing near Madeira island from 20^0 W to the coast of Agadir, where it links with a line of epicenters that follows the trend of the Atlas el Khebir mountains. A triangular aseismic region is thus defined, which includes the Moroccan plateau. From 12^0 W to the east the situation becomes more complicated. There is a critical region around 36^0 N, 12^0 W, where three active lines converge, a continuation of the Monchite fault which runs south-west from Caceres to cape San Vicente, a south-west continuation of the Guadalquivir fault and a line trending south-east towards the Moroccan coast. This has been the epicentral area of large historical earthquakes, such as those of 1356, 1531, 1722, 1755, some of which are reported to have produced large tsunamis on the Spanish and Moroccan coast, and the more recent shock of February 28, 1969, $M = 7$.

East of Gibraltar, seismic activity is spread between a northern limit, that follows the Guadalquivir line and its continuation to the Mediterranean coast near Valencia, and a southern limit along the Riff mountains. Activity in Spain is limited on the east by the Alhama-Carboneras fault and its southern continuation into the Alboran basin. On the coast of Morocco, earthquakes continue eastwards along the trend of the Tell Atlas mountains, through the coast of Algeria and Tunisia with two concentrations near Oran and Orleanville (Girardin et al., 1977). Seismicity on this part is of moderate magnitude, but large destructive earthquakes have happened in the past. Depth is not well known and although some shocks may be as deep as 100 km, most earthquakes are of shallow depth with the exception of the two deep shocks (h = 630 km) in southern Spain (March 25, 1954 and January 30, 1973).

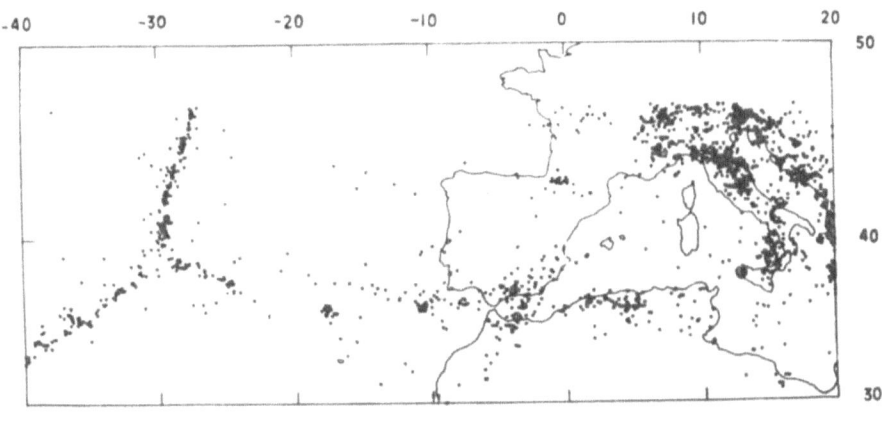

EARTHQUAKES — 1910 – 1977 (NGSDC / EDS / NOAA)

Fig. 1. Epicenter distribution for the period 1910—1977 taken from the Earthquake Data File of the NGSDC, EDS, NOA. All magnitudes are shown with the same symbol

From the north coast of Tunisia, seismicity links with the activity in the Sicily-Calabrian arc, with large earthquakes (1783, 1857, 1905, 1947) and an important concentration of deep foci. Deep shocks are located at depths between 100 km and 450 km at the concave part of the arc. There are two concentrations of activity at depths around 100 km and between 200 km and 350 km with a discontinuity between 100 km and 200 km, and some shocks at 450 km. The seismic zone dips around 60 degrees in WWN direction (Ritsema, 1972). Seismic activity along the Apennines is of moderate magnitude and trends north-west. At the Alps the epicenters are spread along a large convex arc that is continuated eastwards to link with the seismic zone along the coast of Yugoslavia.

From the distribution of epicenters several aseismic blocks may be outlined. Two are formed by the oceanic parts of the Eurasian and African plates and limited to the west by the mid-Atlantic ridge. To the south of the Azores-Gibraltar fault, one may consider the zone formed by the Moroccan plateau and its extension into the Atlantic, outlined by a weak seismic line trending south-east on the Atlantic part and the Atlas el Khabir mountains in the continent north-east. The stable part of Iberian peninsula is outlined to the west and north-west by a weak seismic zone along the coasts of Portugal and north Spain, and the more active zones of the Pyrenees to the north-east and the Betica to the south. Between Spain, northern Africa and Italy there is a large aseismic zone formed by the Balearic, Ligurian and Tyrhenian basins.

A. Udías:

Focal Mechanism and Stress Patterns

A selection of fault plane solutions has been made from those existing for the region with emphasis on those shocks located along the plate contact from the Azores islands to the Calabrian arc. Shocks are listed in Table 1 and their fault plane solutions represented in Fig. 2. Quality of the

Table 1

Identification, number	Date	Latitude ^{0}N	Longitude ^{0}E	Magnitude	Reference
1	17. 9. 1964	44.5	−31.3	5.6	C
2	29. 9. 1965	45.2	−28.2	5.4	C
3	13. 5. 1972	45.0	−28.2	5.0	C
4	11. 7. 1964	41.7	−29.9	4.8	C
5	18. 9. 1964	39.8	−29.7	5.5	C
6	17. 5. 1964	35.2	−35.9	5.6	C
7	18. 11. 1970	35.1	−35.7	5.4	C
8	6. 9. 1964	38.3	−26.6	4.9	C
9	5. 7. 1966	37.6	−24.7	5.1	C
10	4. 7. 1966	37.5	−24.7	5.4	C, A
11	8. 5. 1939	37.4	−23.9	7.1	C
12	25. 11. 1941	37.4	−19.0	8.3	C, A
13	20. 5. 1931	37.4	−15.9	7.1	C
14	29. 6. 1965	36.6	−12.3	4.8	C
15	6. 9. 1969	36.9	−11.9	5.7	C, A
16	28. 2. 1969	36.23	−10.48	5.7	A
17	28. 2. 1969	36.1	−10.6	7.3	C, A
18	15. 3. 1964	36.2	−7.6	6.2	C, A
19	5. 12. 1960	35.6	−6.5	6.5	C
20	2. 7. 1972	36.0	−4.6	4.0	B
21	7. 8. 1975	36.4	−4.4	5.1	B
22	13. 6. 1974	36.88	−4.06	4.2	B
23	30. 1. 1973	37.0	−3.6	4.0	C
24	19. 5. 1951	38.1	−3.7	6.0	C
25	29. 3. 1954	37.0	−3.3	7.2	C, A
26	22. 11. 1972	36.02	−4.07	4.4	B
27	14. 7. 1974	35.58	−3.68	4.3	B
28	17. 4. 1968	35.24	−3.73	5.0	B
29	7. 4. 1970	34.87	−3.90	4.8	B
30	29. 4. 1973	34.63	−4.17	4.5	B
31	24. 8. 1973	35.92	−1.83	3.9	B
32	13. 8. 1967	43.2	−0.5	5.3	A
33	13. 7. 1967	35.5	−0.1	5.0	A
35	9. 9. 1954	36.25	1.53	6.7	A
36	7. 11. 1959	36.4	2.4	5.5	B
37	23. 4. 1967	36.26	2.44	4.8	B
38	24. 11. 1973	36.16	4.40	4.9	B
39	25. 11. 1973	36.14	4.47	4.7	B
40	1. 1. 1965	35.7	4.4	5.2	A, B
41	24. 11. 1973	36.06	4.47	4.9	B
42	29. 6. 1974	36.52	5.21	4.7	B
43	1. 12. 1970	36.7	10.17	4.9	B, BB
44	14. 1. 1968	37.7	13.1	6.0	D

Table 1. Continued

Identification, number	Date	Latitude °N	Longitude °E	Magnitude	Reference
45	25. 1. 1968	37.8	13.2	5.1	A
46	15. 1. 1968	37.9	13.1	5.4	A
47	19. 7. 1963	43.4	8.2	5.5	A
48	23. 12. 1959	37.8	14.7	5.0	D
49	31. 10. 1967	37.8	14.7	4.8	D
50	21. 3. 1972	35.7	15.1	4.5	D
51	16. 6. 1968	37.8	14.8	4.8	D
52	7. 9. 1967	37.9	15.2	4.4	D
53	15. 8. 1967	38.6	15.2	4.4	D
54	23. 11. 1954	38.5	14.9	5.9	D
55	3. 1. 1960	39.2	15.2	6.2	D
56	13. 4. 1938	39.3	15.2	6.9	D
57	28. 12. 1908	38.0	15.5	7.0	D
58	8. 9. 1905	38.8	16.1	6.8	D
59	13. 4. 1973	38.9	17.0	4.4	D
60	21. 8. 1962	41.4	15.5	6.0	A
61	6. 6. 1977	37.38	−1.48	4.2	F

References: A ..... McKenzie, 1972
B ..... Hatzfeld, 1978
BB .... Girardin et al., 1977
C ..... Udias et al., 1976
D .... Riuscetti and Schick, 1975
F ..... Mezcua, 1980

solutions varies from case to case, and readers are referred to the original references for discussion of individual solutions. Fig. 3 shows the horizontal component of the tension and pressure axes. If the horizontal component of one of the axes is much smaller than the other, this is not shown. Finally, in Fig. 4, the horizontal components of the slip vectors are shown

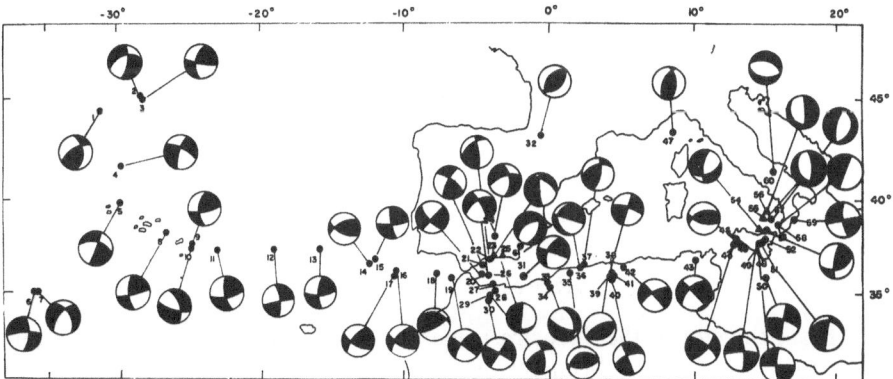

Fig. 2. Fault plane solutions for earthquakes in the area. Numbers refer to those of Table 1. Circles represent the projection of the lower hemisphere of the focal sphere. Compression quadrants are shown in black and dilatation in white

for surface shocks. Only the slip vector of one of the planes is drawn. Selection is based on the interpretation given in this paper to the relative motion of the plates involved. Arrows represent the relative motion of the plate where they are drawn.

Mid-Atlantic and Terceira Ridges

On the mid-Atlantic and Terceira ridges, solutions correspond either to normal or transform faulting. Normal faulting is predominant in the mechanism of shocks 2, 4, 7 and 10, with tension axes perpendicular to the trend of the ridge (Fig. 3). Earthquakes with transform faulting are 3, 8 and 9 with right lateral motion and 6 with left lateral motion. The resulting relative motion at both sides of the northern branch of the ridge, as shown in Fig. 4, is towards the east for the Eurasian and toward the west for the American plate.

On the Terceira ridge, tension axes are also normal to the trend of the ridge, but most slip vectors for the Eurasian plate are in a true eastern direction. This predominant east direction for the motion near this branch

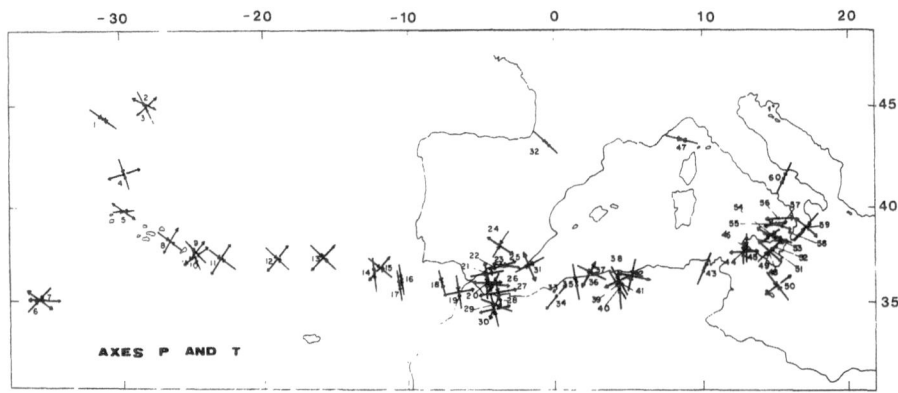

Fig. 3. Horizontal projection of the pressure and tension axes. When the projection of one of the axes is too small it is not shown

of the ridge zone may be explained by a more rapid spreading velocity on the north branch of the ridge than on the Terceira branch. The motion is in both cases consistent with the model of spreading and generation of oceanic crust at the mid-Atlantic ridge.

Azores-Gibraltar Fault

From the end of the Terceira ridge to the strait of Gibraltar runs a long seismic feature that has been recognized as a trancurrent fault (Laughton et al., 1972; Udías et al., 1976). Along this fault, as was already noticed, two parts can be distinguished. From the end of the Terceira ridge to 12⁰ W, mechanism of earthquakes (11, 12, 13 and 15) are consistently of strike-slip right-lateral type with one of the planes striking in east-west direction. This corresponds to pure horizontal motion with the Eurasian plate sliding eastwards with respect to the African plate (Fig. 4). On this zone homogeneous oceanic crust is only involved, which may explain the release of strain by large shocks and the simplicity of motion.

From the east of 12° W to the strait of Gibraltar, thrusting mechanism is predominant (shocks 14, 16, 17, 18 and 19) with consistent horizontal pressure in north-south direction (Fig. 3). Of the two possible planes of faulting, the one representing underthrust of the African plate has been selected (McKenzie, 1972). The slip vectors drawn in Fig. 4 represent the

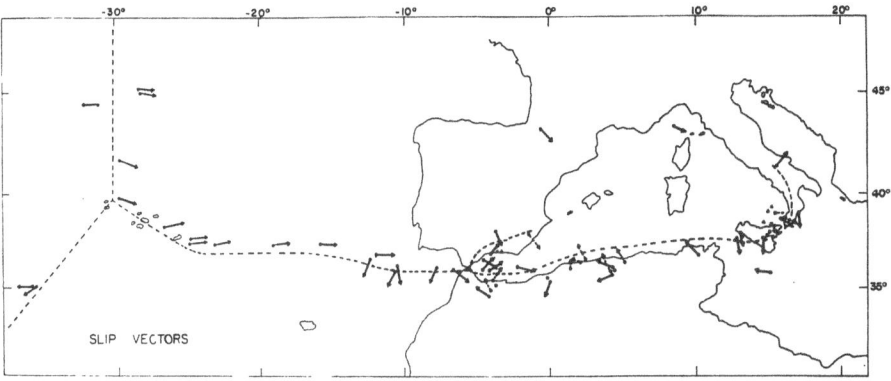

Fig. 4. Horizontal projection of the slip vector. Motion corresponds to that of the plate where the epicenter is drawn

overthrusting motion of the Eurasian plate. The change of motion in this part of the fault may be explained by the interaction of the partly independent Iberian block, which in the present regime produces a locking mechanism against the African block by a counterclockwise rotation. This motion produces an overthrusting of the northern block on the south margin. Another interpretation is that of Purdy (1975), which relates motion in this area to a slow consumption of oceanic crust responsible for the formation of the Gorringe ridge. As was mentioned before, there is sufficient seismic evidence along the west coast of Portugal and the north coast of Spain, to suggest a partly separated block formed by the stable part of the peninsula. Independently of the interpretation given here, in this area, predominant stresses are of compressional nature in north-south direction.

From Gibraltar Strait to Tunisia

East of the strait, epicenters are spread over a wide area including the Betica region, the Alboran basin, Riff area and the north coast of Algeria and Tunisia. The available focal mechanisms show a considerable scatter, due in most cases to the lack of data, because of the low magnitude of the shocks and the poor distribution of stations. On the Riff side, shock 29 shows reverse faulting, consistent with the already mentioned underthrusting of the African plate. Shock 30 is of strike-slip right-lateral motion. In northern Algeria the underthrusting of Africa is present in the mechanism of shocks 35, 37, 39 and 43. Other types of mechanism are strike-slip along east-west faults with either right-lateral (30, 38 and 40) or left-lateral (39) motion.

The situation in Alboran is less clear with different kinds of mechanism (20, 21 and 26). In the Betica region in south Spain, there is a group of shocks (22, 24 and 61) which are of the normal fault type with horizontal tensions. Shock 61 is located in the Alhama fault and it is well documented (Mezcua, 1980).

Stress directions show (Fig. 3) for the north African border a predominant horizontal north-south orientation of pressure axes, consistent with the situation west of the strait of Gibraltar. Certain right lateral motion along east-west trending faults may also be present. However, as most shocks are of low magnitude ($M < 5$), their mechanism may be related to local structures and do not reflect the main tectonic stress directions. This is also the case for the shocks in the Alboran basin and Betica region. In the Betica, a distensive character may be present, with horizontal tension axes. The predominant sense of motion (Fig. 4) is underthrusting of the African plate and a certain eastward component in Alboran and Beticas respect to Africa.

The two deep earthquakes (23 and 25), near Granada in south Spain, have both very similar solutions with a near vertical plane striking north-south and the other horizontal. The pressure axis trends to the west plunging 45 degrees. This solution is not consistent with a simple subduction of the oceanic part of the African plate under Spain to the north. The west trend of the pressure axis seems to relate this mechanism with the Gibraltar arc which is oriented north-south and it is convex to the west. However, the mechanism of these two deep shocks is difficult to relate to existing structures. It may be a relic from a past situation difficult to infer from present conditions.

Calabrian Arc

The Calabrian arc limits to the east the western Mediterranean zone. Main characteristic of this region is the deep seismic activity, which has been interpreted as a subduction zone with a slab dipping 60 degrees to the west, from the convex side of the arc (Ritsema, 1969; McKenzie, 1972; Papazachos, 1973). Mechanism of deep shocks 54, 55 and 56 indicate a pressure axis trending west and plunging in the direction of the sinking slab. Surface shocks 44, 46 and 53 show reverse faulting with pressure axes normal to the arc. Some solutions show strike-slip motion on either north-south or east-west striking vertical planes (48, 49, 50 and 59). As the arc is oriented in a general north-south direction, motion along the east-west striking planes could correspond to faults displacing horizontally the arc. For shocks 51 and 58 the nearly horizontal plane must be selected as the fault plane, representing very shallow underthrusting motion of the crust from the eastern Mediterranean.

The general sense of motion in this subduction zone is, for the shallow shocks, underthrusting of the easthern oceanic crust under the convex part of the arc with some horizontal displacements on vertical faults striking east-west. Mechanism of deep shocks is consistent with a dipping slab under pressure.

Conclusions

The study of the seismicity and focal mechanism data for the region from the Azores islands to Italy indicates a complex behaviour in the western part of the contact between the Eurasian and African plates. The western end of this contact is of ridge nature with horizontal tensions normal to the ridge. From 23^0 W to 12^0 W runs the east-west trending Azores-Gibraltar fault with strike-slip right-lateral motion and characterized by the occurrence of large earthquakes ($M > 7$). From 12^0 W to the strait of Gibraltar the situation becomes more complicated, with the interaction of the Iberian peninsula, which acts as a somewhat independent block, and the convergence of several fractures lines which prolong into the peninsula and the Riff zone. Around 10^0 W there is a critical region, where large earthquakes are generated, such as the famous Lisbon earthquake of November 1, 1755. Stresses in this area are horizontal pressure in north-south direction. Predominant motion is underthrusting of the African plate. East of Gibraltar arc the underthrusting of Africa continues along the northern coast of Morocco, Algeria and Tunisia. A continuous moderate magnitude seismic activity is spread through the Riff, Alboran and Betica region to the northern limit of the Guadalquivir line, with stresses related to local structure, some of distensive nature. This area is also the site of abnormal deep activity (h = 630 km) that may be related to a relic of a paleosubduction zone somewhat related to the Gibraltar arc. At the Calabrian arc a subduction zone is present, with the east Mediterranean crust dipping WWN under the convex side of the arc with the sinking slab under pressure. A western Mediterranean subplate may be defined which includes the Alboran, Balearic, Ligurian and Tyrhenian basins surrounded by the seismic active zones of Spain, northern Africa and Italy and the arcs of Gibraltar and Calabria.

Acknowledgments

The author wants to thank E. B u f o r n and J. S a n c h e z for their co-operation in the compilation of data and drafting the figures.

References

Girardin, N., Hatzfeld, D., Guiraud, R.: La sismicité du Nord de l'Algérie. C. R. somm. Soc. Geol. Fr. fasc. 2, 95—100 (1977).

Hatzfeld, D.: Etude Sismotectonique de la Zone de Collision Ibero-Maghrebine. Doct. Dissert. I. R. I. G. M., Univ. Scien. et Med. de Grenoble. 1—281 (1978).

Isacks, B., Oliver, J., Sykes, L. R.: Seismology and the New Global Tectonics. J. Geoph. Res. *73*, 5855—5899 (1968).

Laughton, A. S., Whitmarsh, R. B., Rusby, J. S. M., Somers, M. L., Revie, J., McCartney, B. S., Nafe, J. E.: A Continuous East-West Fault of the Azores-Gibraltar Ridge. Nature *237*, 217—220 (1972).

McKenzie, D. P.: Active Tectonics of the Mediterranean Region. Geoph. J. R. Astr. Soc. *30*, 109—185 (1972).

Mezcua, J.: Tectonic Implications of the June 6, 1977 Earthquake in Lorca, Spain. Submitted to Bull. Seis. Soc. Am. (in press) (1980).

Papazachos, B. C.: Distribution of Seismic Foci in the Mediterranean and Surrounding Area and its Tectonic Implication. Geoph. J. R. Astr. Soc. *33*, 421—430 (1973).

Payo, G.: Crust-Mantle Velocity in the Iberian Peninsula and Tectonic Implications of the Seismicity in this Area. Geoph. J. R. Astr. Soc. *30*, 85—99 (1972).

Purdy, G. M.: The Eastern End of the Azores-Gibraltar Plate Boundary. Geoph. J. R. Astr. Soc. *43*, 973—1000 (1975).

Ritsema, A. R.: Seismic Data of the West Mediterranean and the Problem of Oceanization. Verhand. K. Ned. Geol. Mijnbouw. Gen. *26*, 105—120 (1969).

Ritsema, A. R.: Deep Earthquakes of the Tyrhenian Sea. Geologie en Mijnbouw. *51*, (5) 541—545 (1972).

Riuscetti, M., Schick, R.: Earthquakes and Tectonics in Southern Italy. Boll. di Geof. *17*, 59—78 (1975).

Udías, A., Lopez-Arroyo, A., Mezcua, J.: Seismotectonics of the Azores-Alboran Region. Tectonophysics *31*, 259—289 (1976).

Address of author: Prof. A. Udías, S. J. Cátedra de Geofísica, Facultad de C. Físicas, Universidad Complutense de Madrid, Madrid 3, Spain.

Rock Mechanics, Suppl. 9, 85—91 (1980)

Rock Mechanics
Felsmechanik
Mécanique des Roches
© by Springer-Verlag 1980

On the Focal Mechanism of Italian Earthquakes[*]

By

C. Gasparini, G. Iannaccone and R. Scarpa

With 4 Figures

Abstract

Compressions and dilatations of all earthquakes occurring in the Italian peninsula have been revised in order to evaluate their focal mechanism and relate then to regional seismotectonic. Some fault plane solutions not reported in the available literature have been determined, including the 1915 Avezzano earthquake ($M_1 = 6.8$).

Focal mechanisms of 28 intermediate and deep Tyrrhenian earthquakes have also been revised. In this case predominant stresses are down dip compressions parallel to the dip of Benioff zone, and extensions axes orthogonal to this, oriented NW-SE.

1. Introduction

The aim of the present paper is to evaluate all fault-plane solutions available for the Italian region, with the exclusion of the Alps. The results from this study are necessary in order to obtain seismotectonic maps and a detailed regionalization of Italy. Data consisting mainly of polarity readings of P-waves have been obtained from many sources, including the BCIS, the ISS and ISC bulletins, and many station readings from WWNSS network and some selected stations. Earthquakes analyzed cover the period 1908—1978 and are concentrated mostly in the Apennines, a region not sufficiently recognized by precedent works on this subject (Cagnetti et al., 1978; McKenzie, 1972; Ritsema, 1969). About 20 new fault-plane solutions of crustal earthquakes have been evaluated and the focal mechanisms of about 30 intermediate and deep earthquakes of the Tyrrhenian Sea have been also estimated. The data are plotted on a Wulff stereographic projection, lower hemisphere. Incidence angles at the focus have been calculated from Herrin et al. (1968) travel-time tables, both from nearby and distant seismic stations. This causes some fluctuations of the data from nearby stations, due to the strong lateral inhomogeneities present in the Italian region. Solutions influenced by this effect and by the uncertainty in the hypocentral depth are certainly those related to the earthquakes of the first half of this century, where more Italian stations are included in the analysis. However the results

---

[*] Pubbl. No. 305, CNR-Progetto Finalizzato Geodinamica — Roma.

0080-3375/80/Suppl. 9/0085/$ 01.40

obtained are not excessively influenced by these factors, as shown by many numerical tests. Solutions have been calculated assuming a double couple as a source mechanism. The computer program used is slightly modified by Wickens and Hodgson (1967), searching numerically the solution which maximizes the function, called score:

$$\text{SCORE} = \left| \frac{\Sigma W_{Pi} \, \text{sgn} \, \varphi_i \, \text{sgn} \, R_i \pm D}{\Sigma |W_{Pi}|} \right|$$

where sgn φ_i and sgn R_i are respectively the theoretical and observed signal polarities, W_{Pi} is a weighting function according to Knopoff (1961). The term D is based on the sign changes between adjacent trial positions and it is introduced on the searching process to emphasize maxima of the score.

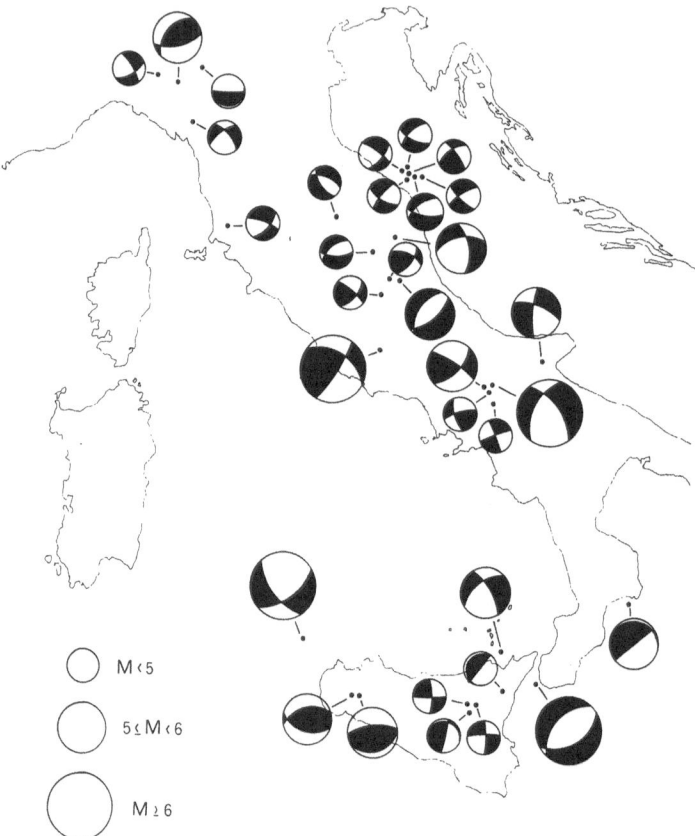

Fig. 1. Fault-plane solutions of crustal earthquakes

In our program, a plot is systematically made of the solutions having the best score, in order to control the main differences between solutions having almost the same number of inconsistent polarity readings. However between the factors affecting the uncertainty of the solutions there are not only errors in the data, but the variability in the number and the inequal space distribution of recording stations, which are particularly few in the lower quadrant. Less reliable are generally the solutions for the earthquakes having magnitudes less than 5. The indetermination introduced by these causes

Fig. 2. Direction of *P* and *T* axes of earthquakes analyzed. The stress component with the most horizontal position is indicated for dip-slip motion; both stresses directions are indicated for transcurrent motion

contribute to a significant reduction in the number of reliable solutions, i. e. those with a possible variation of nodal planes less than 10⁰. Most of the solutions obtained have a possible fluctuation of individual planes amounting to 20⁰—30⁰.

2. Fault-Plane Solutions and Stress Pattern

a) Crustal Earthquakes

About 20 new fault-plane solutions of crustal earthquakes occurring in the Apennines and Calabria-Sicily region are represented in Fig. 1. This map includes some solutions already available in literature (f. i. Riuscetti and Schick, 1975; Cagnetti et al., 1978; Schick, 1978). Earthquakes have been divided into three classes, according to their magnitudes. The events analyzed do not have a homogeneous spatial distribution, but are concentrated in four regions.

In Southern Calabria and Sicily, there is a variety of stress pattern as is illustrated in Fig. 2. Stresses are compressive in Western Sicily, oriented in N−S direction, and tensile in the Messina Strait, oriented in NW−SE direction. This last mechanism, related to the earthquake of December 28, 1908 ($M = 7.0$), indicates normal faulting, aligned with the Comiso-S. Eufemia fault-line. Also the earthquake occurring on March 16, 1941 ($M = 6.9$) near Ustica island, indicates normal faulting oriented along the same direction. Strike-slip motion is predominant in Northern Sicily, showing left-lateral dislocation oriented E − W. The fault-plane solution of the Patti earthquake ($M_L = 5.6$), occurring on April 15, 1978, indicates strike-slip motion with one of the two fault planes aligned with the Eolie-Ibleo-Maltese slope system, characterized by right-lateral motion in this same direction, as shown by recent neotectonic movements (Barbano et al., 1978).

In the Southern Apennines and Gargano region both transcurrent motion and normal faults occur. Tensional axes are aligned in the Anti-Apennines direction. This stress pattern is compatible with the tectonic movements as indicated by other geological and geophysical data (De Vivo et al., 1979).

One of the most interesting new fault-plane solutions is related to the January 13, 1915 ($M = 6.8$), which was one of the most destructive earthquakes occurring this century. The solution is based on clear polarity readings of first arrivals, at 20 seismic stations. This solution indicates transcurrent motion with fault planes oriented in the Apennines and Anti-Apennines direction, wiht left-lateral dislocation in N − E direction. Most of the earthquakes of Central Italy reveal predominant normal faulting and minor strike-slip motion. The tensional axes have a complex pattern, with direction scattered from the principal Apenninic tectonic system. In the Ancona area, six earthquakes occurring during February 1972 show that strike-slip motion is predominant. In this case the fault-planes oriented NE − SW are aligned with some fault systems extending on the land and to the space distribution of some aftershock sequences (Crescenti et al., 1978).

In the Northern Apennines a gradual variation from tensional mechanism types to compressional ones seems to be present. An important transcurrent movement is also present. Recently for the earthquakes occurring in the Tyrrhenian slope of the belt, the plane oriented NNW has been correlated to surface structures of the Tyrrhenian side of the Appenninic belt (Eva et al., 1978).

b) Intermediate and Deep Earthquakes of Tyrrhenian Sea

Fault-plane solutions analyzed are included in the time interval 1938—1978 (Fig. 3). In general, only a few solutions are reliable, both on account of the low number of readings, generally varying between 20 and 40, and the percentage of inconsistent readings.

One of the most reliable and interesting fault-plane solutions is relative to the earthquake occurring on April 13, 1938 (m_B = 7.0), which indicates

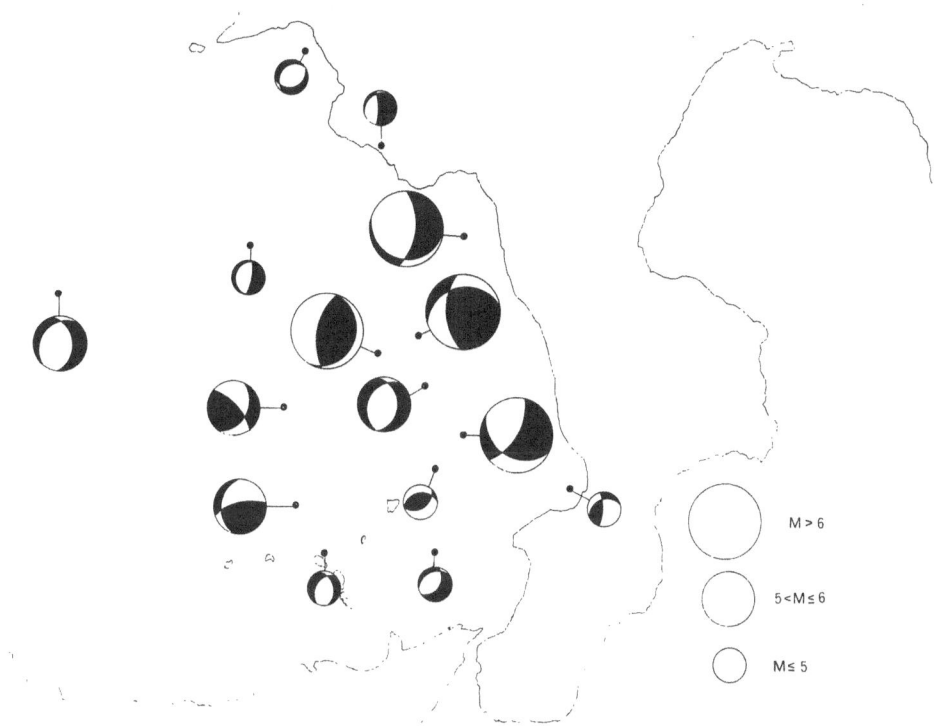

Fig. 3. Fault-plane solution of intermediate and deep earthquakes of Tyrrhenian Sea

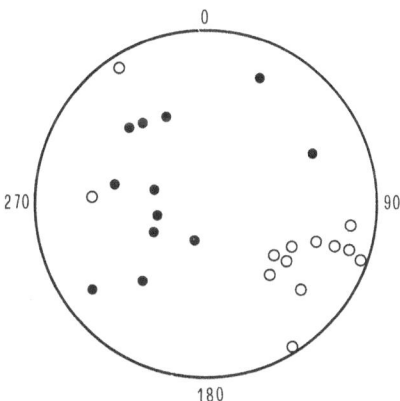

Fig. 4. Stereographic Wulff projection, lower hemisphere, indicating P and T axes of intermediate earthquakes of Tyrrhenian Sea (full and empty symbols respectively)

a pressure axis parallel to the dipping of Benioff zone. Other earthquakes having higher magnitudes have a similar stress pattern. Dip-slip motion is predominant, and the compressional axes have an almost vertical dipping everywhere (Fig. 4). Tensile axes are less scattered than the compressive, and are concentrated in the SE–NW direction in a subhorizontal plane.

An attempt has been made at a regionalization according to the focal depths and the location of these earthquakes, but no simple relationship with other geophysical features seems to be evident.

3. Conclusions

The limits of the present study are due mainly to the low reliability of many fault-plane solutions analyzed. The construction of a more detailed three-dimensional velocity model which is still in progress, will allow more accurate focal depths to be determined and the effects of strong lateral inhomogeneities existing in the Italian region to be taken into account. The results obtained here do, however, form a necessary part of the work of constructing a more detailed seismotectonic models of the Italian peninsula, based on all the geophysical and geological data available.

References

Barbano, M. S., Carrozzo, M. T., Carvani, P., Cosentino, M., Fonte, G., Ghisetti, F., Lanzafame, G., Lombardo, G., Patané, G., Riuscetti, M., Tortorici, L., Vezzani, L.: Elementi per una carta sismotettonica della Sicilia e della Calabria Meridionale. Congr. Soc. Geol. It., Perugia (1978).

Cagnetti, V., Pasquale, V., Polinari, S.: Fault-plane Solutions and Stress Regime in Italy and Adjacent Regions. Tectonophysics 46, 239—250 (1978).

Crescenti, U., Nanni, T., Rampoldi, R., Stucchi, M.: Ancona: considerazioni sismo-tettoniche. Boll. Geofis. Teor. Appl. 73/74, 33—48 (1978).

De Vivo, B., Dietrich, D., Guerra, I., Iannaccone, G., Luongo, G., Scandone, P., Scarpa, R., Turco, E.: Carta sismotettonica preliminare dell' Appennino Meridionale. CNR-Prog. Final. Geodinam. Pubbl. 166, 1—64 (1979).

Eva, C., Giglia, G., Graziano, F., Merlanti, F.: Seismicity and its Relation with Surface Structures in the North-Western Apennines. Boll. Geofis. Teor. Appl. 79, 263—277 (1978).

Herrin, E. (Ed.), Arnold, E. P., Bolt, B. A., Clawson, G. E., Engdahl, E. R., Freedman, H. W., Gordon, D. W., Hales, A. L., Lobdell, J. L., Nuttli, O., Romney, C., Taggart, J., Tucker, W.: Seismological Tables for P Phases. Bull. Seism. Soc. Am. 58, 1193—1352 (1968).

Knopoff, L.: Analytical Calculation of the Fault-plane Problem. Pub. Dom. Obs., Ottawa 24, 309—315 (1961).

McKenzie, D.: Active Tectonics of the Mediterranean Region. Geophys. J. R. astr. Soc. 30, 109—185 (1972).

Ritsema, A. R.: Seismotectonic Implications of a Review of European Earthquake Mechanisms. Geol. Runds. 59, 36—56 (1969).

Riuscetti, M., Schick, R.: Earthquakes and Tectonics in Southern Italy. Boll. Geofis. Teor. Appl. 17, 59—78 (1975).

Schick, R.: Seismotectonic Survey of the Central Mediterranean. In: Alps, Apennines, Hellenides, Closs, H., Roeder, D., Schmidt, K. (Eds.). E. Schweizerbart'sche Verlagsbuchhandlung. 335—338 (1978).

Wickens, A. J., Hodgson, J. H.: Computer Re-evaluation of Earthquake Mechanism Solutions. Pub. Dom. Obs., Ottawa *33*, 1—535 (1967).

Address of authors: C. Gasparini, Istituto Nazionale di Geofisica, Via R. Bonghi 11/b, Roma; G. Iannaccone and R. Scarpa, Osservatorio Vesuviano, 80056 Ercolano (NA), Italy.

Rock Mechanics, Suppl. 9, 93—107 (1980)

Rock Mechanics
Felsmechanik
Mécanique des Roches
© by Springer-Verlag 1980

Theme 3

Geomorphic and Geological Effects of Stresses

Geomorphological Problems in the Alps

By

E. Gerber

With 12 Figures

Abstract

Surface forms are caused by boundary-phenomena of the Earth's crust. In this connection, one has to distinguish between solid and loose rock, as well as between rock-walls and debris-covered slopes. Exogenic as well as endogenic processes are implicated in the fluvial and glacial formation of valleys and ridges; this is also the case in connection with the backwards weathering of wall-slopes. A selection principle can be deduced from the frequency of certain forms.

1. Introduction

The writer is himself primarily a geographer and therefore mostly interested in obtaining a phenomenological description of the surface features as seen today and of the processes that occur thereon.

Phenomenological investigations are for the geophysicist the preliminaries to his studies. The fact that they are being dealt with here, results from a cooperation of 10 years with Prof. A. E. Scheidegger. The cooperation, based on different view-points, has resulted in great mutual benefits, bearing witness to the fruitfulness of interdisciplinary efforts.

2. Surface Problems

Surface forms are boundary phenomena between the atmosphere and the lithosphere, the hydrosphere taking an intermediate position. Within the lithosphere, there exist boundary conditions near the surface which differ from those in the interior. Thus, one distinguishes between *exogenic* and *endogenic* phenomena.

In order to explain surface-features, one always has to take both, endogenic and exogenic processes, into account. Without endogenic processes, there are no differences in altitude; endogenic processes also trigger weather-

0080-3375/80/Suppl. 9/0093/$ 03.00

ing and erosion, for faulting and jointing not only facilitates erosion and predesigns its direction, but also influences weathering and frost-action.

The stresses in the rocks depend also on the external form of the peaks. During erosion, new surfaces with new boundary-stresses are constantly being produced. Thus, in order to explain any geomorphological phenomena, one has to take *three* points of view into account: The outer *form*, the *material* and the type of *process* which is acting near the surface.

3. The Material

3.1 Solid and Loose Rock

For geomorphology, a grouping of the rocks into solid and loose rocks is of fundamental importance. Whilst loose rocks form only a very insignificant part of the lithosphere as a whole, they play a most important part on the Earth's surface.

3.2 The Behavior of Loose and Solid Rock on a Slope

According to the slope-inclination, we can identify *two* types of slopes which are closely related to the properties of the material and which behave fundamentally differently. These are (i) potential debris slopes and (ii) cliff slopes.

The term "potential debris slope" signifies a slope on which loose rock can accumulate without either sliding off on its own or being easily washed off by rain. The maximum slope angle of cohesionless loose rock (the "angle of repose") lies, according to the type of the material, between 30^0 and 40^0. These limits are only rarely exceeded. Since the base, at the mentioned slope angles, is not always covered by debris, we speak of *potential* debris slopes. Non-covered potential debris slopes occur mostly upon limestone, since this type of material is water soluble and produces little residual material.

In general, most potential debris slopes in the Alps are also *actually* covered by debris.

The most characteristic property of the debris-material is its property of being able to absorb water. The water can circulate through the pores, disperse subterraneously or be stored in the ground. Fine material changes its state by the absorption of water. Clay, for instance, passes reversibly through all states from solid through plastic to liquid.

Many different processes producing equilibrium occur in debris slopes, according to slope-surface, material and water content. Thus, one observes mass movements such as creeping, sliding and slumping, erosion and transport of debris.

The *cliff slopes* are in contrast to the debris slopes. Cliff slopes are so steep that no cohesionless rock debris can accumulate on them. If loose material is formed by weathering or through the action of stresses, it slides or glides quickly off the cliff-slope. Water runs over such slopes without any erosive action; brooks that enter the cliff from above form waterfalls, wash the rock but erode only upon impact at the bottom which may cause fluting and undercutting of the wall. Fluvial erosion can occur on cliffs only if the latter form the lateral walls of a gorge. The same is true with regard to glacial erosion which can only occur when the ice flows along the cliff under pressure.

One has to draw a distinction between cliffs consisting of coherent loose rock and cliffs consisting of solid rock. Cliffs built of coherent loose rock seldom attain a height of more than 10 m; they may, if they become soaked with water, suddenly collapse. They will not be discussed here any further. Cliffs built of solid rock can absorb only little water and do not change their state thereby. They are stable even when great precipitation occurs.

Since free-standing rock-cliffs do not decay under humid and moving debris covers, since erosion is basically absent on them and since they consist usually of selectively exposed resistant rock-types, their recession is very slow. Their decay is caused not only by exogenic erosion but also by stress concentrations in the rock which induce jointing.

4. Surface Forms and Structures

In order to explain the form of cliff slopes, one has to start with the material. One distinguishes between homogeneous (not layered, not banked, as commonly observed in crystalline areas) and layered rock of alternating resistance to erosion (as commonly observed in sedimentary areas). First we

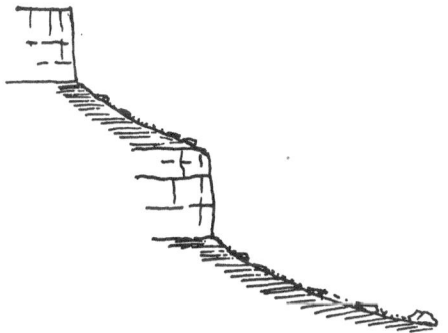

Fig. 1. Horizontal layering. Alternation of debris-covered and cliff-type slopes. Canyon-type

shall ignore the jointing. The simplest arrangement is horizontal layering (Fig. 1). In a rock-wall, an alternating sequence of layers of different resistance to erosion leads to a stepped slope, in which cliff-slopes alternate with debris-covered slopes. The most impressive example of this kind is the Grand Canyon of the Colorado River in the U. S. A.; one could, thus, designate this type of arrangement as "Canyon-type". If the layers are inclined, three possibilities exist: the *layer-slope*, the *head-slope* and the *cross-slope* (Fig. 2).

It is easy to realize that head-slopes and cross-slopes lead to forms of cliffs which differ but little from those obtained in horizontal layering. With regard to the layering, such cliffs are stable. Slope recession occurs by means of break-offs; sliding is absent.

Phenomenologically the most varied are layer-slopes. In this case, one can identify three types: The *overcut* layer slope (slope is less steep than the layers), the *parallel* layer slope, and the *undercut* layer slope (slope is steeper than the layers).

The overcut layer slope is the most stable type. In parallel layer slopes, the slope surface is the layer-surface itself. If the dip angle β of the layer is large enough so as not to be able to support loose debris, the slope becomes a cliff slope which, according to the actual value of the dip angle, may not be much steeper than a debris slope. Weathered material is able to slide off leading to rock slides.

The undercut layer slope is the least stable type. Here, the layer-surfaces may even act as slide-horizons. Examples of large slides of this type are those of Goldau in 1806 (Heim, 1932) and Vajont in 1963 (Müller, 1968).

We have indicated that the term "erosion" (particularly of rock cliffs) encompasses quite different phenomena. This may indicate the formation of a very thin patina of a dark color at the surface, so that recent break-offs

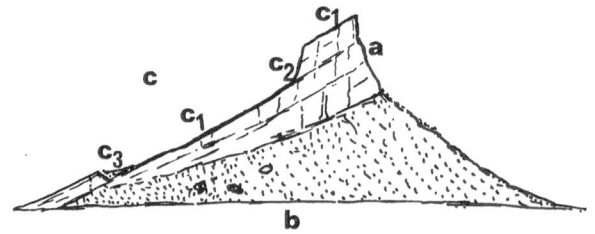

Fig. 2. Inclined layers. (a) Head slope, (b) Cross slope, (c) Layer slope, (c$_1$) Slope parallel to layers, (c$_2$) Undercut slope, (c$_3$) Overcut slope

can be recognized by their lighter color. Deeper chemical weathering may lead to volume changes and fracturing. Stress concentrations can cause platey exfoliation parallel to the surface. Fractures and joints provide places for frost action; this is of great significance in the Alps where temperature fluctuations around the freezing point are common. This action should not be overestimated regarding its significance in connection with cliff recession.

5. Phenomenology of Rock Stresses

No rocks of any size exist without showing jointing and fracturing. In fact, the joints can be so closely spaced that the rock, tectonically, behaves as a "loose" material. All transitions between widely to very narrowly spaced joints are in existence.

Joints and fractures are seldom homogeneously distributed. In general, there are preferred orientations and it is one of the main tasks of Alpine geomorphology to determine these. The joints are caused by stresses which are due to two types of causes: First, by the effect of neotectonism and second by the action of surface-dependent form-induced effects (which includes self-gravitational stresses). It is not always possible to separate these two causes unequivocally.

Typical form-induced stresses exist in a cliff if the rock at the foot is incompetent owing to lithology (Fig. 3). In this case, tension-cracks arise at the upper bounding plane of the cliff which are parallel and orthogonal to

Fig. 3. Limestone cliff
I Unstressed cliff zone with protrusion, *II* Zone with free recession, *III* Base

the wall (Gerber and Scheidegger, 1965). If the rock falls off, the cliff recedes in parallel fashion. In a wall, the stress increases from top to bottom because of the increasing overburden. In horizontal layering, the upper layers may stick out. At the lower parts of the wall, with increasing load, wall-parallel joint surfaces are created. One can identify three zones on a cliff: (i) the unstressed top, (ii) a zone of free weathering in which the decaying rock is held in place by the stresses and the recession is thereby impeded, (iii) a foot-zone showing increased jointing. At the top we have a tensional, at the bottom a compressional stress state.

It is a most striking fact that most high cliffs (i. e. some 100 m) show plane-parallel recession so that extensive cliff-zones may be formed. The planar recession cannot be explained by exogenic weathering alone. It has to be the expression of an extensive stress field (Brunner and Scheidegger, 1973). If one inspects such cliffs (Fig. 4), one recognizes mainly steep and vertical fracture surfaces which are parallel and orthogonal to the wall. In addition, one observes conjugate vertical fracture surfaces which form an angle of about 45⁰ with the wall, and finally inclined-conjugate surfaces whose intersection line is horizontal and normal to the wall. The vertical

fracture surfaces produce the side-walls of couloirs, the inclined ones provide
the basis for the formation of peaks.

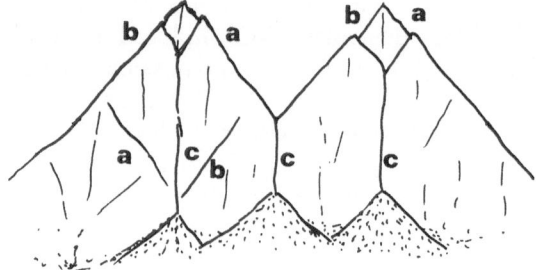

Fig. 4. Fracture surfaces in a cliff and peak formation. The main fracture surface is the
front-surface; (a, b) conjugate fractures, (c) vertical fracture

Let us note that, during recession, the cliff does not become less steep
as has been proposed in certain models (Fig. 5 a) (Bakker and Le Heux,
1947). There would be no steep cliffs in the Alps if this were the case, for

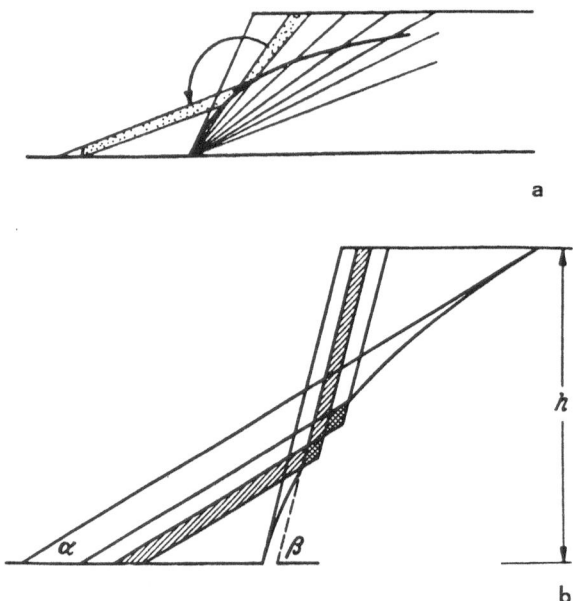

Fig. 5. Cliff recession (a) by progressive decrease of the declivity (wrong!), (b) parallel cliff
recession and rock-core formation beneath the scree slope

all cliffs are subject to slope recession. On the other hand, an inclined scree
slope, i. e. a type of debris-covered slope, is formed at the foot of the cliff
(Fig. 5 b), by which the height of the cliff is progressively reduced if the
debris is not constantly removed.

6. The Genesis of Valleys and Peaks

In the preceding section, existing cliffs that recede have been discussed. The process of cliff recession can be observed by many details. The opposite phenomenon, the genesis of valleys and peaks, is much more difficult to observe. In fact, the primeval design of the large Alpine valleys is shrouded in the dark of the past and the trend of the "primary valleys" of the Alps is subject to much speculation (Staub, 1934).

On an inclined surface, one can identify two types of river nets:

(i) River nets in which *freely branching* gulleys combine to form branches and finally a trunk (Fig. 6). A comparison with a "tree", albeit only in two dimensions, suggests itself. Such a river-net is called a freely developing one. The small rivulets enter the bigger branches and the latter enter the trunk at an acute angle φ which is called the *branching angle*. Every river net can be regarded as a drainage area. In the present case, the latter has the form of a leaf narrowing at the mouth.

(ii) River nets in which the valley trends are remarkably straight; the lateral valleys enter the main valleys at a branching angle $\varphi \sim 90^0$ (Fig. 7). Even the branches that form the trunk are usually symmetrically positioned. The consistency and symmetry of the pattern reminds one of an artificially trained tree (along a wire grid); for this reason, we call this type of pattern "trellis-pattern".

If one investigates the connection of the two above types of patterns with the underlying ground, one notices that freely-developing patterns occur mainly in loose rock and in shallow valleys. These conditions may be

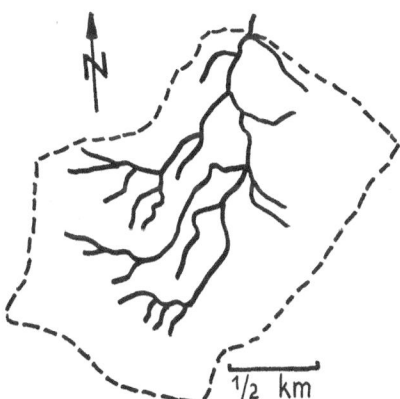

Fig. 6. Freely branching drainage, schist in flysch. Tributary of the Waldemme, Switzerland

found on a large scale in sedimentary plains and on a small scale on debris-covered slopes (Fig. 6). River nets of the trellis-type are usually found cut into solid rock (Fig. 7).

If one directs one's attention to the development of Alpine water-courses of valleys, one has to make conjectures regarding the "primeval surface". The large thrust-sheets (nappes) exposed to-day are extended slide horizons. Therefore the thrust sheets are represented as sweeping, curved, well-defined lines on large-sclae geological cross-sections. This interpretation has been

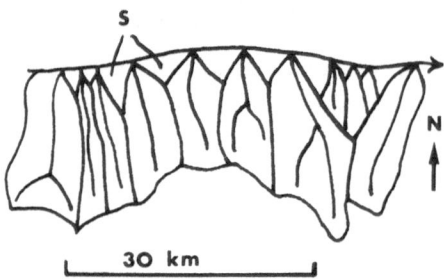

Fig. 7. Trellis pattern. S sectors; longitudinal Salzach valley, Austria

applied in many places and the hypothetical connections across eroded regions are probably justified.

However, regarding the smoothness of the uppermost nappe-surface, along which no other nappes were thrust, it is probably justified to entertain some doubts regarding the above interpretation. One can imagine that here, on the uppermost unstressed rocks, the surface was much torn and highly jointed, because one has to assume that gravitational slumps of torn-off "packets" must have occurred, even if the thrust surfaces were inclined. If this idea is accepted, then it is clear that the "primeval" valleys could not have developed freely on a slightly undulating surface, but had to follow tectonically predesigned lines. One can also assume that displacements and new movements occurred in the later phases after the valley trends were originally designed. This does not contradict the contention that incised valleys are tied to the original design during further erosion. This corresponds to the contention of those geologists who maintain that the present-day valleys often do not follow the lithological boundaries, but possibly older ones (Heim, 1922).

At any rate, leaving the above speculations aside, we shall proceed now to a discussion of the phenomenology of present-day valleys.

In the last century, the Alpine valleys were considered as huge cracks which were created during the genesis of the mountain ranges in question. (Heer, 1879). Erosion was held to be of but little importance. In Switzerland, under the influence of Heim, the above idea was generally abandoned. Heim (1921) maintained that valleys are caused solely be fluvial erosion and that the result of this process is a mountain-region without any tectonic influence in the phenomenology of its large-scale features. The notion of "pure erosion", however, was usually applied very loosely, for this term is an incomplete one, since there cannot exist erosion "as such"; the term must always refer to a certain material. The resulting features bear out the characteristic traits of the *type* of erosion (fluvial, glacial) as well as of the *material* (crystalline, limestone, friable, slates).

In fluvial erosion, the river "erodes" only the valley floor directly. The formation of the slopes occurs depending on the inclination and on the type of rock, by erosion in the slope-gulleys and by mass-movements which, on cliffs, are represented by the fall-off of loose debris. The valley-river thus, does not simply only "erode", but it also removes the material originating from the valley-sides. If the power of transportation is insufficient, the valley floor is raised by the progressive accumulation of material. Our attention must be focussed on the sides, on which debriscovered slopes as well as cliff slopes may occur. This may occur either in an extreme fashion where the side is formed by but one type of slope, or in a mixed fashion where both types of slopes may alternate with each other.

In the Alps, the case may occur where, depending on the type of rock, the valley sides may be covered from top (water-shed) to bottom (valley floor) by debris of scree. The various processes occurring on such debris slopes (creeping, sliding, slumping, wild-water formation) are only mentioned here in passing, as our attention is to be focussed on cliffs.

In the Alps, no valleys exist whose side would be formed of cliff slopes all the way from the water-shed down to the valley floor. Narrow gorges are very rare as well. The usual cases are valleys in which cliff-slopes and debris-covered slopes occur concurrently; the debris-covered slopes are predominant; they occur even in the highest regions.

The recession of the cliff-slopes depends on their position in the altitude-scheme. One has (i) cliffs at the bottom of a valley-side, (ii) cliffs between debris-covered slopes and (iii) cliffs in the watershed-region with peaks.

Let us first consider cliff-slopes in the valley region.

Owing to fluvial erosion, a gorge is formed in solid rock. If one travels through a gorge, one often observes so-called rock-mills in the lower parts of the cliffs which were hollowed out by the vortex-motion of debris-laden water. Equally, one observes plane parts on the walls which correspond to joint-surfaces. In the higher regions which have already been subject to recession, typical exfoliations predominate which are independent of the direct erosion of the river in the valley. However, it is not necessary that cliffs are the result of receding gorge-walls. For cliffs to be created, it suffices that the valley-sides be undercut by lateral erosion. Once cliffs have arisen, there is only an indirect effect by the river on the recession by the removal of the fallen debris.

During the recession, great influence is exerted by the different types of material involved. In layers of little resistance to erosion, debris-covered slopes arise above, below and between the cliffs. This leads to a stepped appearance of the valley-side (Fig. 1).

For many years, the view was current (Annaheim, 1946) that many flat parts of a valley side are remnants of old valley floors. Countless geomorphological investigations (see e. g. Uhlig, 1954) confined themselves to attempting reconstructions of the old valley floors. However, such a monogenetic viewpoint is unable to explain the polymorphism of the valley sides. Old valley floors are completely obliterated by the recession-process so that they can no longer be recognized. This fact is even more evident if the glacial, in addition to the fluvial, action is taken into account.

7. Glacial Effects on Valleys: Glacial Troughs

The effect of glacial action on Alpine valleys is a very impressive phenomenon. Many valleys have been widened; they have broad valley floors and steep sides often in the form of cliffs. Traces of the sliding of the ice (striations) are often visible. The distinction between fluvial and glacial effects in valleys has usually been based on the assumption that rivers cause a V-shape and glaciers a U-shape. This, however, is only conditionally true, inasmuch as purely fluvial valleys exist that have a wide floor with steep sides (U-shape); similarly glacial valley reaches exist with a typically V-character.

The insufficiency of the mentioned assumption is also evident in models that have been advanced as scientifically sound. Usually, such models are purely formalistic. The oldest of them is the model of a glacial trough (Fig. 8) which was widely discussed around 1912 (see e. g. Penck, 1912)

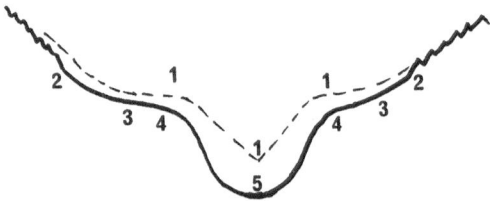

Fig. 8. Schematic cross section through trough-shaped valleys (obsolete)
1: Possible preglacial profile, 2: Ice-cut hollow, 3: Ice-planed surface, 4: Shoulder of trough,
5: Wider or deeper part (floor) of the trough

and is current in many textbooks of to-day (see e. g. Louis, 1968). It should be noted, however, that ice is not a "sculptor" who wants to fashion a trough. For steep cliffs to arise, it is necessary that the rock be solid and that fracture surfaces exist which are parallel to the valley. Ice is then a suitable agent for exposing the fracture (joint-)surfaces. Matthes (1930) stated with regard to the Yosemite Valley in California: "... where the rock is plentifully divided by natural partings, the glacies will quarry out entire blocks and excavate at a fairly rapid rate." What Matthes stressed for the Yosemite Valley, is also true e. g. for the classical "trough" of the Lauterbrunnen Valley in Switzerland (Fig. 9). A purely formal true-scale cross-section is not sufficient to explain the profile, since it is the type of rock and the fracture (joint) pattern which influences the shape; the wide valley-floor is the result of the damming effect of a post glacial mountain slide (Gerber, 1945).

8. The Patterns of Catchment Areas

The drainage pattern in the Alps is characterized by large longitudinal valleys which are fed by extensive catchment areas. Except for areas whose drainage is subterraneous (as is the case in karst areas of the Calcareous Alps), one can delineate specific catchment areas exhaust the map in a com-

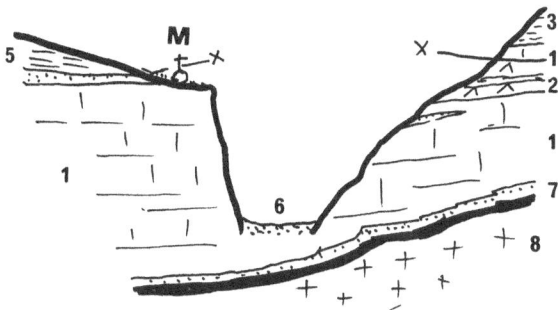

Fig. 9. The classical trough of the Lauterbrunnen Valley in Switzerland
Geological cross section
M: Mürren, 1: Compact limestone, 2: Limestone with marl layers, 3: Marl, 4: Sandstone, 5: Slate, 6: Valley-bottom, accumulation behind a landslide, 7: Limestone with marl layers, 8: Crystalline

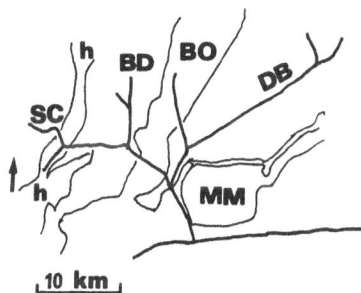

Fig. 10. Candelabrum-pattern of a river system. Valpelline, Aosta Valley, Italy. The valley-system ranges across several nappes
DB: Dent Blanche, BO: Grison-schist, BD: Bernhard thrust sheet, MM: Mont Mary thrust sheet, SC: Zone Sion-Courmayeur, hh: Zone houillière

plete pattern. The pattern differs from a random mosaic by the fact that, owing to the declivity from the water-shed to the main valley, the catchment areas are pointed in the latter direction.

Owing to the complete coverage of the region by catchment areas, not all of them can be convex everywhere: convex, leaf-shaped domains cannot completely cover an area. The most frequently observed type is that of pointed strip (Fig. 7) which is drained by a simple single valley. Between the points, triangular rest-surfaces are left which drain directly towards the main valley; they are nothing but the left-over, cut-up parts of the original

sides of the main valley; we call them "sectors". All these types of symmetrical valley patterns, of which Fig. 10 shows the Valpelline as an example,
and which include the characteristic "arched" valleys, cannot be "random"
occurrences, but indicate the presence of a tectonic design which is remarkably independent of lithological boundaries.

9. Forms of Watershed Ledges

Alpine valleys are but rarely so narrow that their sides form a single
sweep from the peaks of the watersheds to the valley floor. In most cases,
there is an intermediate high region in form of a more or less wide ridge or
plateau from which the peaks which form the water-shed, stick out. The
erosion in such intermediate regions is largely independent of the deeply-
cut valleys themselves.

On the ledges, peaks exist which remind one of constructed buildings.
This impression is based on more than a superficial similarity. Free-standing peaks are only possible because they are statically well designed. Fluvial
erosion plays only a minor role in their genesis, since not much water can
be collected near the water sheds, because it falls off the steep cliffs and
becomes dispersed. Water can play a role in transport only in rock-fall gul-

Fig. 11. Trihedral peak with ledges as supports
1: Ledge, 2: Additional Ledge

leys (Fig. 4). Ice can freeze to a wall, but it cannot move there. Similarly,
the frost-action should not be overestimated. The resulting features are mostly
determined by the properties of the rock, the layering therein and, especially,
by the jointing. Generally, cliff-surfaces recede by stress-induced joint-surfaces.

Many features are built in exactly the same way as if they had been
designed by elastostatic calculation. Thus, in a high cliff, the inclination
decreases from top to bottom in homogeneous rock. The cliff on a peak
between two edges is often concave. The edges are stressed; the intervening
wall is statically more stable if it is concave than if it were straight. Very
often, it is possible to observe a break-out below a summit. If a stress-
bearing edge is undercut, two additional edges are formed which provide
the static support (Fig. 11).

10. The Selection Principle

During the action of the processes of erosion and weathering, particularly resistant rocks are left over and form protrusions, steps, terraces and ribs. These features are thought to be the result of selective erosion. Furthermore, this "selection" can be regarded as a principle which is active quite generally in the formation of the Earth's surface features and which refers to the static conditions (Gerber, 1969).

A protrusion in a cliff is subject to shearing forces, by which it may be brought to a break-off. On the other hand, a part which conforms to the static stress field, may be kept in place even if it disintegrates. If a piece breaks off a wall, one can usually observe that a supporting arch is formed.

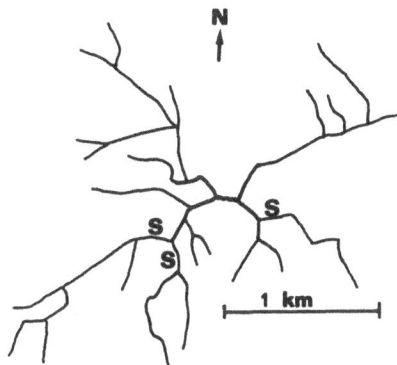

Fig. 12. Matterhorn with support-ledges (S)

If rock breaks off between two supporting edges in a glaciated area, and if ice can accumulate in the resulting niche, the debris material caused by further weathering will slide off the ice or be removed by it; thus no debris-cone can be formed. The stress concentrations lead to further break-offs until a statically stable hollow is created.

Ice-filled hollows at the foot of a mountain peak are called *glacial cirques*. Their genesis can be explained only on the basis of the assumption that the stress conditions and the most stable configuration in the rock are considered along with the processes of erosion and removal of material by ice. On steep cliffs, the erosive action of ice is minimal, because the ice tends to flow away. For the genesis of a cliff, one has to invoke mainly the stress-conditions as significant. Once the statically most stable convex form has been attained, further recession is slow. The ice acts solely as a transporting agent, for, in ice-free cirques, scree slopes are formed at the foot of cliffs leading to a covering-up of the rock. The erosive action of the ice is most pronounced at the foot of the cliff and on the floor of the cirque, where ice flows across the substratum under pressure. The stress-maximum occurs immediately below the foot of the cliff, where the ice on the slope increases the load. In special cases, a hollow cirque may be created which turns into a lake upon melting of the ice. Thus, a "glacial" cirque is not caused solely by the "erosive" action of the ice; rather, *it is a remnant peak in a statically preferred form.*

From the type of observations adduced above one may deduce a general *selection principle* which is valid for all surface-forms:

The processes causing surface-changes proceed the more slowly, the more stable the surface-form already present is. Statically unstable forms decay quickly and tend to develop into statically more stable forms.

Thus, there is a *selection of statically well-designed forms.*

The Alpine peaks often appear as bold structures. Thus, the Matterhorn is (Fig. 12), as a static structure, very stable and well balanced. Large parts of it, which are prone to weathering, have long since decayed. The high cliffs which are exposed to the action of a rough climate, decay on their surface, leading to rock falls which, however, do not change the large-scale shape.

Most ledge-shapes and protruding peaks have developed into statically stable structures by continuing selection.

Even in the best-designed shapes, stress redistributions arise on occasion by continuing weathering and by fluvial and glacial erosion. This occurs mostly at the foot of the slope which leads to the formation of joints. Thus, unstable parts may be formed which decay rapidly in conformity with the selection principle. These processes take place during the course of millenniums. In the final stage, a transition from cliff-slopes to debriscovered slopes occurs entailing entirely different recession processes; provided that no new altitude differences are created by tectonic effects which would trigger new erosion and new stress-induced cliff shapes.

Acknowledgments

The work reported here was carried out in cooperation with the Institute of Geophysics of the Technical University of Vienna, under the auspicies of the Austrian Geodynamics Program. Field work during several summer seasons in Austria was financed by the Austrian Geodynamics Committee; this support is gratefully acknowledged.

References

Annaheim, H.: Studien zur Geomorphogenese der Südalpen zwischen St. Gotthard und Alpenrand. Geogr. Helv. *1*, 65—149 (1946).

Bakker, J. P., Le Heux, J. W. N.: Theory on Central Rectilinear Recession of Slopes. Konin. Nederl. Akad. van Wettensch. V. *50*, 959—1162 (1947).

Brunner, F., Scheidegger, A. E.: Exfoliation. Rock Mech. *5*, 43—62 (1973).

Gerber, E.: Lage und Gliederung des Lauterbrunnentales und seiner Fortsetzung bis zum Brienzersee. Mitt. Aarg. Natf. Ges. *22*, 165—184 (1945).

Gerber, E.: Bildung und Formen von Gratgipfeln und Felswänden in den Alpen. Z. Geomorph. Suppl. *8*, 94—118 (1969).

Gerber, E., Scheidegger, A. E.: Probleme der Wandrückwitterung, im besonderen die Ausbildung Mohrscher Bruchflächen. Felsmech. und Ingenieurgeol. Suppl. 2, 80—87 (1965).

Heer, O.: Die Urwelt der Schweiz. 2. Aufl., 713 pp. Zürich: Schulthess (1879).

Heim, A.: Geologie der Schweiz. 3 Vols. 704/1018 pp. Leipzig: Tauchnitz (1919/1922).

Heim, A.: Bergsturz und Menschenleben. 218 p., Zürich: Fretz und Wasmuth (1932).

Louis, H.: Allgemeine Geomorphologie. 3. Aufl., 522 pp. Berlin (West): Walter de Gruyter & Co. (1968).

Matthes, F.: Geologic History of the Yosemite Valley. Washington: U. S. Gov't Printing Office (Geological Survey) Prof. paper 160. 137 p. (1930).

Müller, L.: New Considerations on the Vaiont Slide. Felsmech. und Ingen. Geol. 6, 1—91 (1968).

Penck, A.: Schliffkehle und Taltrog. Petermanns Geogr. Mitt. 1912 (2), 125—127 (1912).

Staub, R.: Grundzüge und Probleme alpiner Morphologie. Denkschr. Schweiz. Natf. Ges. 69, Abh. 1, 183 p. Fretz Zürich (1934).

Uhlig, H.: Die Altformen des Wettersteingebirges mit Vergleichen in den Allgäuer und Lechtaler Alpen. Forsch. Deutsch. Landeskunde 79, 1—103 (Remagen: Verlag der B. A. für Landeskunde) (1954).

Address of author: Dr. E. Gerber, Krummenland 307, CH-5107 Schinznach-Dorf (AG), Switzerland.

Rock Mechanics, Suppl. 9, 109—124 (1980)

Rock Mechanics
Felsmechanik
Mécanique des Roches
© by Springer-Verlag 1980

Alpine Joints and Valleys
in the Light of the Neotectonic Stress Field

By

A. E. Scheidegger

With 11 Figures

Abstract

It is contended that the orientation of joints as well as of valleys in the Alps is intrinsically determined by the orientation of the neotectonic stress field in that region. Conversely, the orientation of the principal neotectonic stresses in the Alps can be inferred from a statistical analysis of the orientation of joints and valleys. Evidence is adduced to support the above contention on a small (local) as well as on a large (continental) scale.

1. Introduction

The present investigation is concerned with two Alpine phenomena: joints and valleys. In fact, these are the two prominent features that determine the physiography of a mountainous area. On a micro-scale, joint planes form the surface of any rock-outcrop. On a macro-scale, valleys (and their complements: the peaks) determine the morphology of the region as a whole.

The forces that influence the shape of the surface of the Earth have two possible origins: either they originate in the interior of the Earth (endogenetic forces) or they originate in the atmosphere (exogenetic forces). The present-day surface morphology of the Earth is the result of the antagonistic action of these two types of forces (principle of antagonism; Scheidegger, 1979 a). Generally, endogenetic forces (due to the tectonic stress field) tend to build up an area, exogenetic ones (weathering, erosion) tend to tear it down. In this, the action of the endogenetic forces is ordered, that of the exogenetic forces, random. The basic character regarding the randomness or non-randomness of a phenomenon allows one to determine whether it is of endogenetic or exogenetic origin.

Furthermore, the action of the two types of forces is, geologically speaking, extremely rapid. Geotectonic rates of rising of 1 mm/year have been commonly observed in the Alps. Corresponding rates of weathering are similar: measurements of material-transport of the major rivers referred to their drainage areas also give values of the same order of magnitude, so that there exists a dynamic equilibrium. Extrapolated, the value of mm/year

0080-3375/80/Suppl. 9/0109/$ 03.20

yields km/million years which shows that the present-day morphology of the Alps cannot be very old, at most a few hundred thousand years: the mountains are constantly being eroded and re-created.

The basic argument of this paper is that both, joints and valleys, have non-random orientations and are therefore caused by endogenetic forces. As these forces, the tectonic stress field is identified. Furthermore, since evidence shows that the surface features in the Alps, which includes joints and valleys, are of very recent origin, it is the *neo*tectonic stress field which, to a large extent, must be responsible for the origination of both types of features. In turn, the nature of the neotectonic stress field (i. e. the orientation of the principal stress directions) can be determined from an analysis of Alpine joints and valleys.

2. Joints

2.1 The Nature of Joints

All rocks in outcrops show the general phenomenon of being jointed. The joints appear phenomenologically as cracks or potential cracks which are exposed by natural weathering. Thus, any recent outcrop generally shows a jagged appearence, the exposed surfaces being indicative of the intrinsic orientation of the joints.

It is clear that not all types of joints have the same origin. For some types, e. g. columnar jointing in lava, an obvious origin by cooling can immediately be assigned. However, most "ordinary" joints in an outcroup do not fall into such an obvious category, and thus there is some question regarding their origin, a subject which has been discussed in detail by S c h e i d -e g g e r (1979 b). Thus, most common joints are planar and smooth as if cut by a knife. They have no fillings, no slickensides and no evidence of any motion at all. Visible heterogeneities, like pebbles in a breccia or "nagelfluh" are commonly cleanly cut through. At a single outcrop, one finds ordinarily three joint systems which are usually very definite: one system is nearly horizontal (dips 0^0 to 40^0) and corresponds to some lithological factor; the other two systems are nearly vertical and almost orthogonal to each other. In any one region, there is generally great conformity between the orientations of the near-vertical joints, with at most 20% of the outcrops showing orientation-anomalies to perhaps 30^0, indicating that their orientation is non-random and therefore that they are of endogenetic origin. Evidence has been mounting that they are slip-surfaces in the neotectonic stress field. As such, they would be expected to be oriented as bisecting surfaces between the largest and smallest principal stress directions containing the intermediate principal stress. This is not too far removed form the idea that joints are Mohr-type fracture surfaces; however, the latter form smaller angles with the maximum compression direction than do the shear surfaces; inasmuch as the evidence shows that conjugate joint sets intersect each other at close to 90^0 and not 60^0, their interpretation as shear planes is preferred (S c h e i d e g g e r, 1979 b).

2.2 Stress Determination from Joints

Since recent joints, on the average, appear to have a definite orientation with regard to the neotectonic stress field, it is possible to deduce the latter from an analysis of the orientation of joints. In the case of Mohr-type fractures, the bisectrix of the *smaller* angle between joint sets at an outcrop should be the greatest compression. However, inasmuch as, as noted, the

Table 1

Location	Number of data	Joint set I	Joint set II	Angle	P	T
Switzerland	20 regions	$74 \pm 13/90 \pm 11$	$346 \pm 15/88 \pm 12$	87	120/1	210/2
Austria	28 regions	$26 \pm 22/89 \pm 12$	$270 \pm 24/90 \pm 13$	64	148/1	238/0
Spain	170	$84 \pm 10/90 \pm 7$	$181 \pm 12/90 \pm 8$	84	133/0	43/0
Southern Alps	402	$1 \pm 7/90 \pm 5$	$272 \pm 8/89 \pm 6$	90	137/1	46/1
Central Yugoslavia (Visegrad)	66	$246 \pm 26/66 \pm 18$	$162 \pm 5/67 \pm 5$	76	114/1	23/30
Macedonia-Olympus	180	$234 \pm 12/82 \pm 12$	$327 \pm 6/75 \pm 5$	89	100/16	191/5
Antalya	44	$203 \pm 17/87 \pm 14$	$123 \pm 16/86 \pm 12$	80	73/0	343/4
Eisenach (DDR)	29	$207 \pm 20/89 \pm 15$	$289 \pm 22/86 \pm 15$	82	158/2	68/3
Dalen (Norway)	88	$263 \pm 21/73 \pm 10$	$204 \pm 5/90 \pm 5$	60	145/16	53/10

angle between steeply dipping conjugate joint sets is usually close to 90^0, it is often not possible to distinguish reliably between the largest and smallest principal stress direction.

The determination of the preferred joint orientations in an area has to be carried out by a statistical procedure. For this purpose, a computational method was developed by Kohlbeck and Scheidegger (1977). In that method, two statistical probability distributions of the type $\exp (k \cos^2 \vartheta)$ about a mean direction are fitted to the data; the two best-fitting mean directions are determined by computer using a function-minimization procedure. The largest (P) and smallest (T) prinicpal stress directions, then, are the bisectrices of the joint sets, P being located in the smaller quadrant.

The above idea has been applied to the joints and stress-field in the Alpine-region. Table 1 gives a summary of the results. In this table, the data are represented as follows. The orientation of a joint set is given by stating its dip direction (azimuth $N \rightarrow E$ in degrees) and dip (angle with the horizontal in degrees) separated by a solidus. A straight line (force) will be represented by giving the azimuth ($N \rightarrow E$ in degrees) of its plunge downward and the plunge angle (angle with the horizontal in degrees), again separated by a solidus. Errors are indicated by $\pm$ and represent the 90% confidence limits. The table lists, in sequence, the area under consideration, the number of joints analyzed, the two near-vertical joint-orientation maxima, the angle between them and the P and T directions. For visualization, the horizontal projections of the principal stress directions are shown as plotted at their respective locations in Fig. 1. In this figure, it should be kept in mind that the identification of P and T is not certain.

2.3 Discussion of Results from Joints

The writer has already collated the results from joint-orientation interpretations with those obtained from other means of stress-field determinations. In general, there is a very good correspondence (Scheidegger, 1979 c).

One may wish to compare the stresses obtained from joint-orientation measurements in the Alps with those elsewhere in Europe. Data from the limestone quarry near Dalen in Norway were made available to the writer.

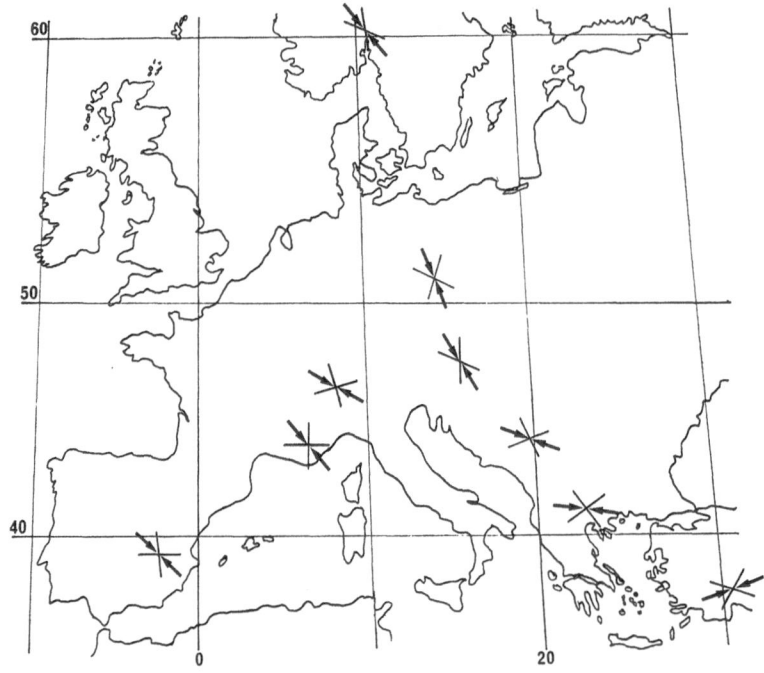

Fig. 1. Plot of preferred joint sets and maximum compression direction in Europe

Similarly, a set of joint-orientation measurements near Eisenach (GDR) was found in the literature (Weber, 1962). These joint-orientation measurements were evaluated using the usual statistical procedure; the results are also shown in Table 1. For better visualization, the results have also been plotted on a map in Fig. 1.

From an inspection of the results it becomes obvious that there is evidently a large-scale European neotectonic stress system present which is homogeneous from the Alps north to Norway. Significant changes occur only to the east, in the Dinarides and in Turkey.

3. Valleys

3.1 The Nature of Valleys

An inspection of the drainage basins on a map of a mountainous area shows that certain typical patterns recur again and again. Such patterns are for instance that of a "candelabrum" and that of an "arched valley". A series of such patterns in nature have been collected and discussed some time ago (Gerber and Scheidegger, 1977) which led to the conjecture that they have been created by the geotectonic stress field. In this instance one can assume that the physical cause of the valley trends and of the joints might be the same. Thus, as for the joints, it will be assumed that the valleys are also discontinuity surfaces in a triaxial stress field. Naturally, the valleys would not be assumed to be rifts or fissures in the ground; it is rather suggested that the general trend of their direction is outlined by the stress field.

It may come somewhat as a surprise that valley trends should be of very recent origin. However, we are dealing in the Alps essentially with a *young* mountain region. In such an area, the rates of raising by geotectonic forces (and a corresponding lowering of the land by erosion so that a dynamic equilibrium is achieved) is, as has been shown in the Introduction, of the order of millimeters per year. In the context of this remark, the contention that the present-day morphology in mountainous area is of very recent origin, is perhaps somewhat more understandable.

A drainage pattern that is of purely erosional origin should be a *random* pattern, so that the orientation structure should not show any pronounced maxima. An attempt to find such a drainage basin showed that even in manifestly random-like patterns, preferential directions are evident, indicating a geological control. Other than the gullying in a heap of mine tailings, it appears that most water courses are influenced by geological control.

3.2 The Statistical Procedure for Drainage Patterns

In order to make a quantitative evaluation of valley structures, it is first of all necessary to define the "direction" of a valley quantitatively. Inasmuch as a river or creek has generally a rather curved, meandering course, the fixation of a "direction" is not a priori possible. It is therefore necessary to start with a given map; on such a map, the "blue lines" are taken as the elements that represent a drainage network. Another map of the same area, particularly if it is on a different scale, would, of course, represent a somewhat different set of elements; in this instance, the very production of the map by the cartographer is, in a way, arbitrary. However, this arbitrariness is at least not biased in the direction of the further statistical analysis.

Inasmuch as the "blue lines" on a map are "wiggly lines" in the sense of Ghosh and Scheidegger (1971) they must be "rectified". The procedure suggested to do this is as follows. The drainage pattern of the map is considered as an arborescence, in which the free and internal vertices are connected by *straight* lines. For the straight lines, it is easy to determine their direction and length. The direction is the azimuth of the line N → E in degrees, the length is measured in some suitable units and is used as the weight in the subsequent statistical evaluation. For the latter, however, we shall use the *normal* to (i. e. the "poles" of) the directions of the links rather than the directions themselves, because there is then a simple correspondence with joint orientations: The links can be considered as traces of vertical joints (dip 90⁰) with the dip direction equal to the normal to the links.

After the graph of a drainage pattern has been rectified and measured as to direction and weight of the links, one can construct the polar histogram from which maxima can be read by a nonparametric procedure. However, it is better to effect the further evaluation by means of the Kohlbeck-Scheidegger (1977) procedure developed for joints. Thus, again a series of Dimroth-Watson (of type exp $(k \cos^2 \vartheta)$) distributions are fitted to the data and the best-fitting parameters thereof are determined by means of a maximum-likelihood procedure. This will yield values for the best fitting azimuths of the basic distributions including confidence limits. If there are

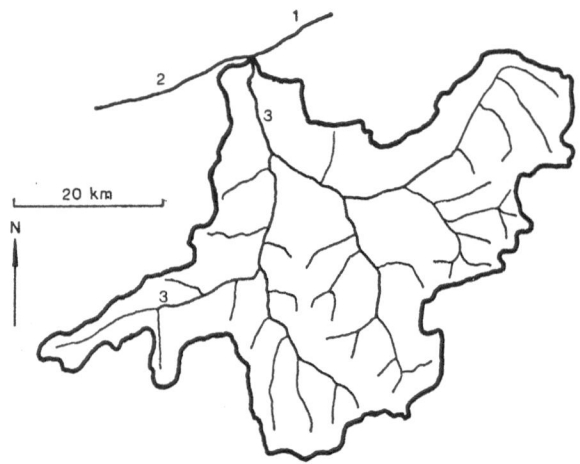

Fig. 2. The candelabrum of the Hinterrhein valley in Switzerland

only two basic distributions, the bisectrices of them can be taken as the principal stress directions under the assumption that the links are lines of maximum shear. If the links are supposed to be Mohr-type fractures, the greatest pressure would be contained in the smaller quadrant. However, which is the latter, is uncertain and the identification as to which of the principal stress directions is the greatest and which is the smallest pressure, is not assured statistically.

3.3 Two Typical Alpine Valley Patterns

3.3.1 *The Candelabrum*

A. Morphology

A frequent type of drainage patterns occurring in mountainous areas is that of the candelabrum. A typical case of this type is the Hinterrhein drainage area in Switzerland. Fig. 2 shows an outline of the pattern.

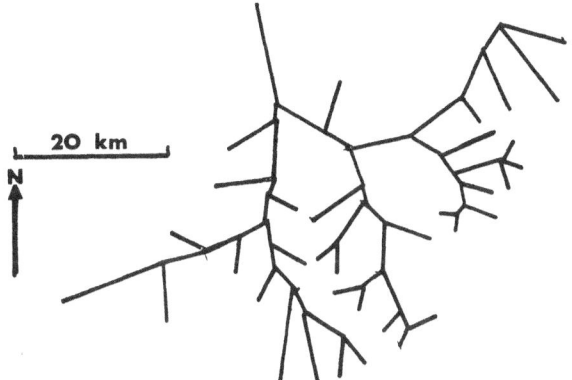

Fig. 3. Rectified candelabrum

Geologically, one finds here the whole section of Alpine rocks from Tertiary to Crystalline.

B. Valley Orientations

The drainage system of the Hinterrhein was "rectified" in the way described earlier. The graph representing the rectified arborescence is shown in Fig. 3. With this graph, a statistical analysis was made of the directions

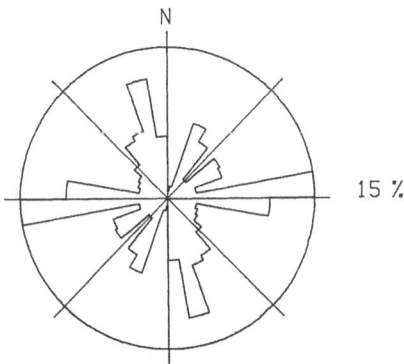

Fig. 4. Polar histogram of the rectified candelabrum

taking as length-unit 1.33 km. The histogram of the direction-poles is shown in Fig. 4. It is at once obvious that there is a simple structure inasmuch as there are two well-defined maxima in the histogram. Indeed, the computer-evaluation according to the Kohlbeck-Scheidegger (1977) procedure yields

Max. 1	72 ± 6	66%
Max. 2	156 ± 3	34%
Angle	84^0	
P	$24^0/0^0$	
T	$114^0/0^0$	

C. Geotectonic Implications

The orientation of the Hinterrhein-Candelabrum fits exactly into the orientation patterns found for Switzerland as a whole (Scheidegger, 1979d). The orientation structure of the joints in Switzerland shows pole maxima at the azimuths of $74^0 \pm 13^0$ and $166^0 \pm 15^0$, which are very close indeed to the azimuths of the Hinterrhein valleys. Accordingly, the direction of the principal stress directions calculated under the *assumption* that joints as well as valleys are shear lines in a tectonic stress field, come out very close to each other, viz. for the joints the azimuth of P is 120^0, for T it is 30^0. This differs by 6^0 from the values for the Hinterrhein valleys; in the light of the indicated error estimates of the statistical estimates, this discrepancy is insignificant.

One difficulty arises with the *identification* of which is the P- and which is the T-direction, though. Generally, one assumes that the P-direction is that principal stress direction which is enclosed by the *smaller* angle between the strikes of the directions. Under this assumption, the P- and T-directions are reversed in the case of the valleys and of the joints. However, it is again found that the angle in question is above 80^0, so that the identification of which is smaller and which is larger than 90^0 is by no means statistically assured. Furthermore, the writer has adduced evidence that the joints (and valleys) are probably not Mohr-type fracture planes, for which the angle-relation would hold, but shear planes in a creep-type movement for which the Mohr-relation regarding angles of intersection does not apply.

Thus, one may conclude that the causes and the strikes of the valley patterns investigated in the present context coincide with those of the joints in the area.

The formation of the candelabrum itself is then simply due to the fact that a drainage pattern *must* form an arborescence; the branches of the latter follow successively the prescribed directions. This produces a candelabrum.

3.3.2 Arched Valley

A. Morphology

The next type of valley which we shall discuss is the arched valley in the classification of Gerber and Scheidegger (1977). As a typical example we choose the drainage basin of the Nassfelder Ache near Badgastein in Austria. Fig. 5 shows the drainage system in question.

The rocks consist mostly of central gneiss of the Tauern Window containing a preponderance of potassium-feldspar. Occasionally, exposures of black phyllites and marbles are also encountered. Morphologically, the dominating feature in the area is the Kreuz-Kogel (indicated by K-K in Fig. 5)

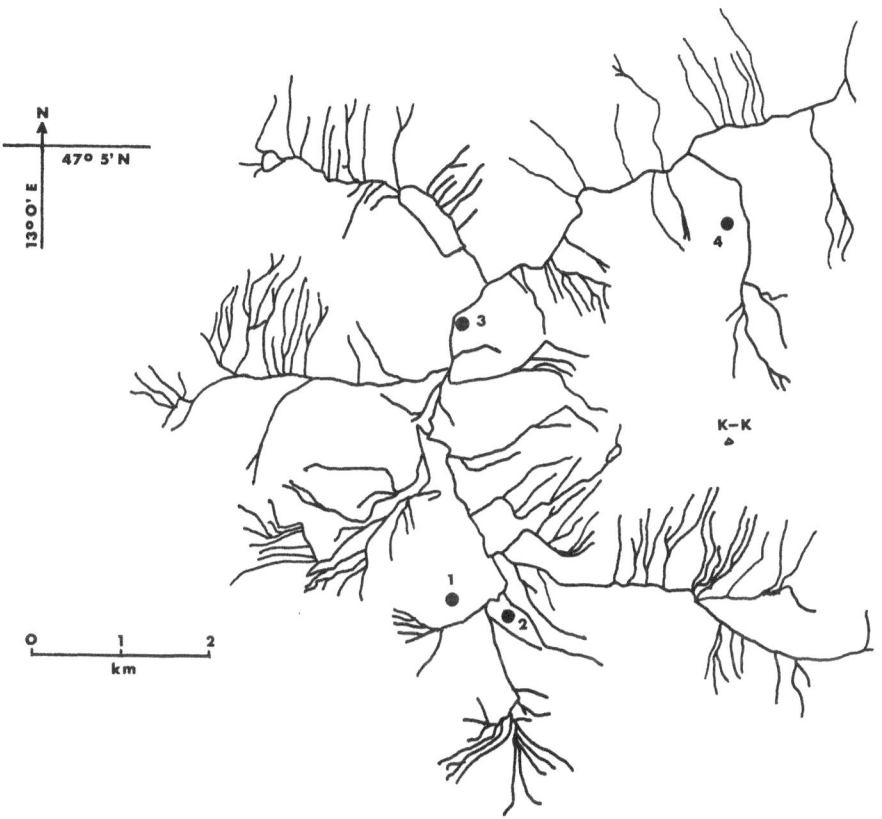

Fig. 5. Drainage system of the Nassfelder Ache near Badgastein, Austria (arched valley)

whose peak reaches an altitude of 2686 m above sea level. This has to be compared with the elevation of 1131 m of the lowest point of the Nassfelder Ache (confluence with the Gasteiner Ache).

B. Valley Orientations

The drainage system of the Gasteiner Ache was "rectified" in the way we described, leading to the graph shown in Fig. 6. This graph was then digitized with length units of 250 m. Fig. 7 shows the polar histogram for the resulting distribution of directions. As usual, not the strike directions, but the corresponding poles have been plotted. It is quite obvious that the distribution of directions is rather poorly defined. Nevertheless, the computer was able to determine two significant maxima whose pole positions are

$$
\begin{array}{llll}
\text{Max. 1} & 76^0 \pm \ 8^0 & \text{weight} & 54\% \\
\text{Max. 2} & 151^0 \pm 10^0 & \text{weight} & 46\%
\end{array}
$$

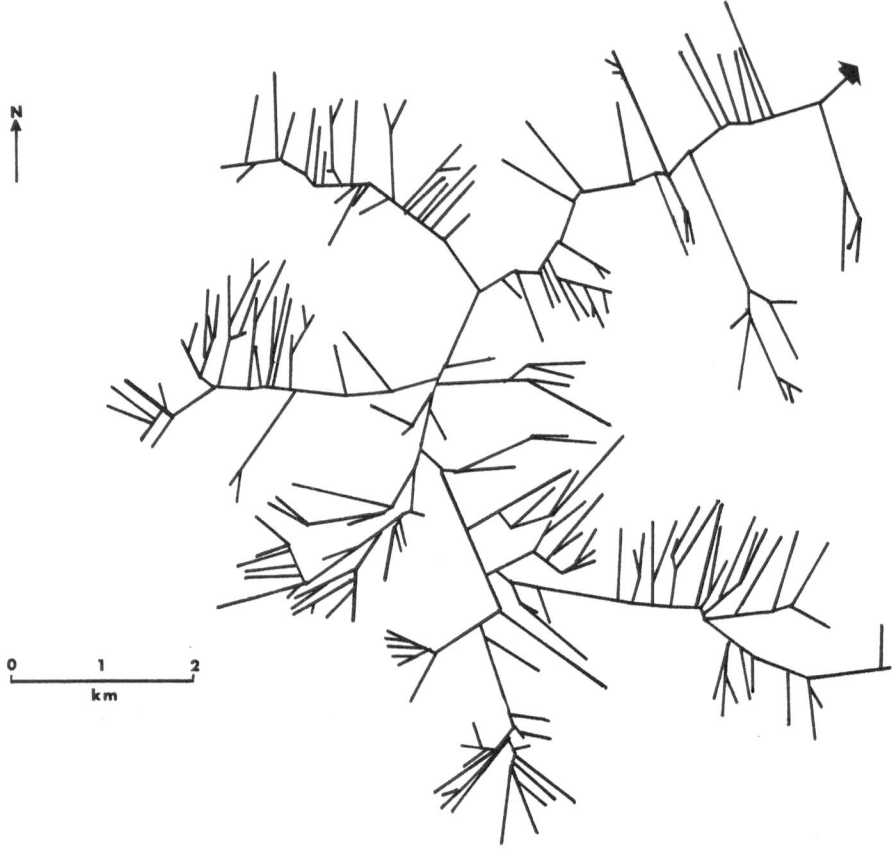

Fig. 6. Rectified arched valley

If the valleys are assumed to be fracture lines of a geotectonic stress system, the orientation of the latter would have to be

$$
\begin{array}{ll}
P & 23^0 \\
T & 113^0
\end{array}
$$

C. Joints

In order to compare the distribution of the orientation of the arched valley with further tectonic data, the orientations of 80 joints were measured at four outcrops in the area whose locations are indicated by the numbers

1—4 in Fig. 5. The combined pole-density diagram for these joints is shown in Fig. 8. It is quite evident that only one maximum of orientation is definite,

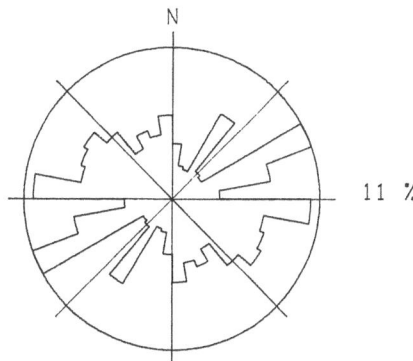

Fig. 7. Polar histogram of arched valley

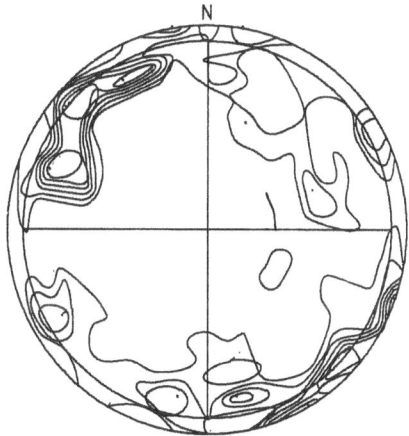

Fig. 8. Pole-density diagram of joints in the Nassfelder Ache region

a fact which is confirmed by the computer-evaluation with yields only one significant maximum, viz.

$$\text{Max.:} \quad 143^0 \pm 16^0/86^0 \pm 11^0$$

It is seen that this agrees very well with the second maximum of the valley directions.

D. Comparison

The above data yield a consistency (within 8^0) between the preferred direction of the joints and at least one of the prevailing valley directions in the area under consideration.

On the other hand, the joints — and valley — orientations in the region of the Nassfelder Ache do not agree with those prevalent in the adjacent regions. We have made a similar analysis for the region of Badgastein (Hauswirth and Scheidegger, 1980) and have found the values shown in Table 2. Again, joint and valley orientations agree mutually (within 9⁰) in the adjacent Badgastein region, but not with the orientations in the drainage area of the arched valley. The discrepancy is from 30^0—50^0, which is significant.

Table 2. Gastein Region

	Max. 1	Max. 2	P	T
Joints	$217 \pm 8/88 \pm 5$	$113 \pm 10/90 \pm 8$	345/2	75/1
Valleys	23 ± 9	122 ± 11	72/0	162/0

It should also be noted that the orientation values for the Gastein region are also those that are typical for the Alps as a whole. The Nassfelder Ache, i. e. the "arched valley", is therefore in every regard and anomaly.

E. Geotectonic Model

It remains to explain the features observed in arched valleys from a mechanical standpoint. The main physiographic feature, as has been men-

Table 3

Deviation angles of principal stress directions from stress direction at infinity in a uniaxially stressed plate with a rigid circular inclusion at the origin. The radius of the inclusion = 1 unit, x is parallel to the stress direction at infinity, y is normal to x

x	$y = 1$	$y = 2$
0	0.0	0.0
0.25	27.8	1.2
0.5	15.3	2.1
0.75	9.2	2.2
1.0	8.0	1.7
1.25	8.1	0.8
1.5	7.6	0.3
1.75	6.9	1.3
2.0	5.9	2.0
2.25	5.0	2.3
2.5	4.2	2.5
2.75	3.6	2.5
3.0	3.0	2.4
3.25	2.5	2.3
3.5	2.2	2.2

tioned above, is the mountainous mass of the Kreuzkogel. This is the center around which the valley forms its "arch". Evidently, the existence of this mass has something to do with the "deflection" of the preferred orientations of the joints and valleys from their "normal" positions.

If the crust of the Earth is considered as a plate under stress, it would be quite expectable that any elastic inhomogeneity would cause an inhomogeneity in the stress field as well. The course of the stress trajectories around a circular hole in a plate is well known (see e. g. Goodier, 1933) the stress trajectories around a circular *rigid* inclusion are similar (Pulpan, 1969); the deviation angles of the principal stress directions from their values at infinity are shown in Table 3. It is seen that values reach almost 30^0; in fact, closer to the inclusion, the values would be even higher.

If now the valley-arch is formed in only one part of the circular "inclusion", it is clear that any mean value up to about 30^0 can be obtained, which could explain the observed values.

3.3.3 The Large-scale Orientation of Alpine Valleys

Thus far, we have considered only small-scale valley patterns and we have been able to show that there is a correspondence between the local trends and the local joints.

However, it stands to reason that a corresponding relation also exists between continental valley-trends and the continental stress field.

Indeed, this problem has already been studied by Scheidegger (1979d, e) for Switzerland and Austria. Accordingly, it was shown that the preferred valley orientations in *Switzerland* are such that the principal stress direction is N 115^0 E, whereas the corresponding stress direction from joints is N 120^0 E. In Austria, the principal stress direction from valley orientations is N 132^0 E, that from joints is N 148^0 E. In both cases, therefore, there is a close correspondence, as has been claimed.

3.3.4 Attempt at Finding a Random Drainage Pattern

It has been noted some time ago (Gerber and Scheidegger, 1973) that "pure erosion" is characterized by an intrinsic randomness. One would expect, therefore, that a river system which is not influenced in any way by tectonic control, would show a random distribution of links. It is not easy to find such a river system inasmuch as even in plains areas, the underlying "basement tectonics" may manifest itself in the orientation of the river links.

After some searching, we felt that the best chance of a "purely erosional" river system might be found in the alluvial flats of the Canadian Arctic. This is an area covered by swampy muskeg and permafrost. The area chosen is that of the Anderson River, as shown in Fig. 9.

We then "rectified" the system of links as described earlier, which led to the graph shown in Fig. 10. The links were measured as to orientations and lengths (in units of 5.8 km). The histogram of the resulting distribution of directions is shown in Fig. 11. An inspection of this figure shows that there are at last 5 maxima which can be identified. This would indicate a random distribution of link directions. However, the computer still can find 2 well defined preferred orientations for the link directions. One obtains

Max. 1	8 ± 8	51%
Max. 2	94 ± 6	49%
Angle	86^0	
P	141/0	
T	51/0	

The computer procedure does not converge consistently for the assumption of 3 or 4 maxima. The distribution of the link directions, therefore, is evidently not really random.

3.4 Discussion of Results for Valleys

In our discussion of various valley patterns we have found that the pattern of a candelabrum fits entirely into the scheme that would be expected if the valleys were fractures induced by the neotectonic stress field. This has

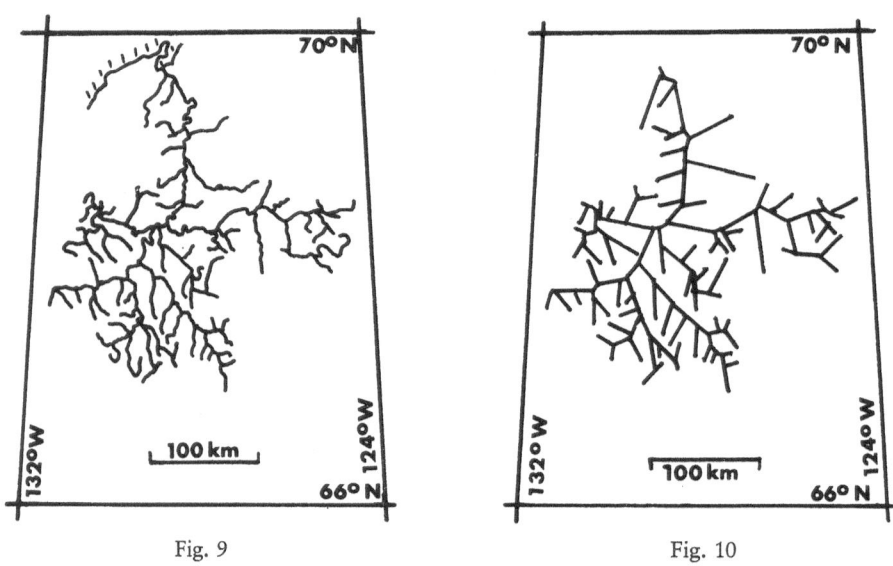

Fig. 9   Fig. 10

Fig. 9. Drainage pattern of the Anderson River (Canadian Arctic)
Fig. 10. Rectified drainage pattern of the Anderson River

also been found to be the case for very large-scale studies of whole regions such as Switzerland and Austria. Even apparent "random" patterns (see the Anderson River basin) are found to have a significant orientation-structure, which is evidently caused by some tectonic control.

However, characteristic smalle-scale features such as "arched valleys" evidently do show some significant deviations from an orientation-structure that would be expected from what is known about the regional tectonic stress field. However, joint and valley orientations still agree with each other, although they do not agree with the large-scale regional trend, show-

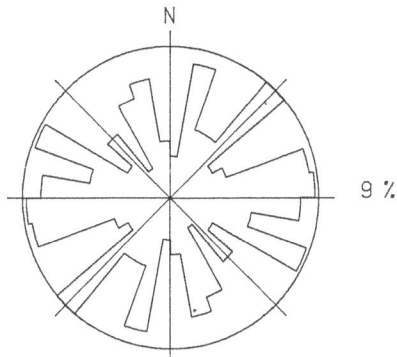

Fig. 11. Polar histogram of Anderson River links

ing that such valleys are the expression of an anomaly of the geotectonic stress field. On a large scale, the valley trends in Austria and Switzerland closely correspond to those of the joints in the corresponding regions.

4. Conclusion

When we review the evidence adduced in the paper, it is noted that the orientations of joints as well as of valleys are non-random and therefore, presumably endogenetically induced. Furthermore, the preferred orientations of joints and valleys agree with each other in any one area. On the whole, these orientations also fit into the continental orientation patterns, although anomalies were found to exist.

The joint and valley orientation patterns can be explained in terms of the orientation of the principal neotectonic stress directions. In Europe, there is a large-scale pattern wherein the maximum compression is NW−SE. Joints and valleys, except in anomalous areas, fit into this pattern. The anomalies, such as the arched valley near Badgastein, can be explained in terms of reasonable geomechanical models.

Acknowledgements

The writer is indebted to Dr. H. P. Geis of the Mining Division, Elkem-Spigerverket a/s, Oslo, for making the joint-orientation data from the limestone quarry at Dalen available to him. The computing center of the Technical University of Vienna provided calculation-facilities at no cost, without which the work reported here could not have been done. The entire project was carried out under the auspices of the Austrian Geodynamic Committee (chairman: Prof. F. Steinhauser) whose financial support is gratefully acknowledged.

References

Gerber, E., Scheidegger, A. E.: Erosional and Stress Induced Features on Steep Slopes. Z. Geomorph. Suppl. Bd. *18*, 38—49 (1973).

Gerber, E., Scheidegger, A. E.: Anordnungsmuster von alpinen Tälern und tektonischen Spannungen. Verh. Geol. B.-A. *1977* (2), 165—188 (1977).

Ghosh, A. K., Scheidegger, A. E.: A Study of Natural Wiggly Lines in Hydrology. J. Hydrol. *13*, 101—126 (1971).

Goodier, J. N.: Concentration of Stress Around Spherical Inclusions and Flaws. Trans. Amer. Soc. Mech. Eng. *55*, 39—44 (1933).

Hauswirth, E. K., Scheidegger, A. E.: Tektonische Vorzeichnung von Hangbewegungen im Raume von Bad Gastein. Interpraevent 1980 (in press) (1980).

Kohlbeck, F., Scheidegger, A. E.: On the Theory of the Evaluation of Joint Orientation Measurements. Rock Mech. *9*, 9—25 (1977).

Pulpan, H.: A Complex Variable Technique for the Stress Field Around an Elliptic Underground Inhomogeneity. Pure and Appl. Geophys. *76*, 137—146 (1969).

Scheidegger, A. E.: The Principle of Antagonism in the Earth's Evolution. Tectonophysics *55*, T7—T10 (1979a).

Scheidegger, A. E.: The Enigma of Jointing. Riv. Geofis. Sci. Aff. *5*, 1—4 (1979b).

Scheidegger, A. E.: The Geotectonic Stress Field and Crustal Movements. Symposium on "Recent Crustal Movements", IUGC Meetings, Canberra Dec. 1979. Tectonophysics, in press (1979c).

Scheidegger, A. E.: Orientationsstruktur der Talanlagen in der Schweiz. Geograph. Helv. *34*, 9—15 (1979d).

Scheidegger, A. E.: Beziehung zwischen der Orientationsstruktur der Talanlagen und den Kluftstellungen in Österreich. Mitt. Österr. Geograph. Ges., *121*, 187—195 (1979e).

Weber, H.: Die Oberflächenformen der Tambacher Schichten bei Eisenach: Klüfte. Forschungen z. Deutschen Landes- und Volkskunde *24* (2), 96—101 (1926).

Address of author: Prof. Dr. A. E. Scheidegger, Institute of Geophysics, Technical University, Gusshausstrasse 27—29, A-1040 Wien, Austria.

Rock Mechanics, Suppl. 9, 125—138 (1980)

Rock Mechanics
Felsmechanik
Mécanique des Roches
© by Springer-Verlag 1980

Tectonic Stresses as Determined from the Character of Fault Systems in the Bohemian Massif

By

Naděžda Šťovíčková

With 6 Figures

Abstract

The main fault systems in the Bohemian Massif are classified on the basis of geophysical indications, space orientation and their correlation with epitectonic, volcanic and metallogenic phenomena. The aspects of depth, dynamic function and mutual relations of faults are taken into account. The interpretation of linear density discontinuities by the Linsser method, computed regional magnetic anomalies and results of deep seismic sounding provide the main basis of the presented results. The theory of rotational dynamics and problems of the historical development, based on the principle of paleoorientation deduced from paleomagnetic data are discussed in brief.

1. Introduction

This paper is the result of several years of interpretation-work of geophysical data from the Bohemian Massif (Pokorný et al., 1975—1979) aiming at classifying and characterizing the main fault systems. Three structural schemes presented here (Röhlich, Šťovíčková, 1968; Blížkovský, Pokorný, Weiss, 1975 and Pokorný, Šťovíčková, 1979 — Figs. 1, 2 and 4) express the development of ideas published first by the author (Vondrová, 1963) and step by step refined by means of interpretation of geophysical (gravimetric, magnetometric, seismic and petrophysical) data. An attempt is made to characterize not only the course of structural discontinuities, but also their depth, dynamic function, morphology and history.

The leading method in interpretation is the application of the Linser method of determining indications of linear density-discontinuities from Bouguer gravity anomalies, i. e. the computer-processing of measured gravity data and the theoretical course of gravity anomalies above a vertical density step. The calculation has been carried out at two depth-levels (-1 and -2 km below the surface). The mutual position of indications from different levels can be used for the determination of the dip of the discontinuity plane. The comparison of residual gravity anomalies with the Bouguer anomaly field

was used as well as the map of horizontal gravity gradients. The results of
this interpretation have been compared with the magnetic field and with

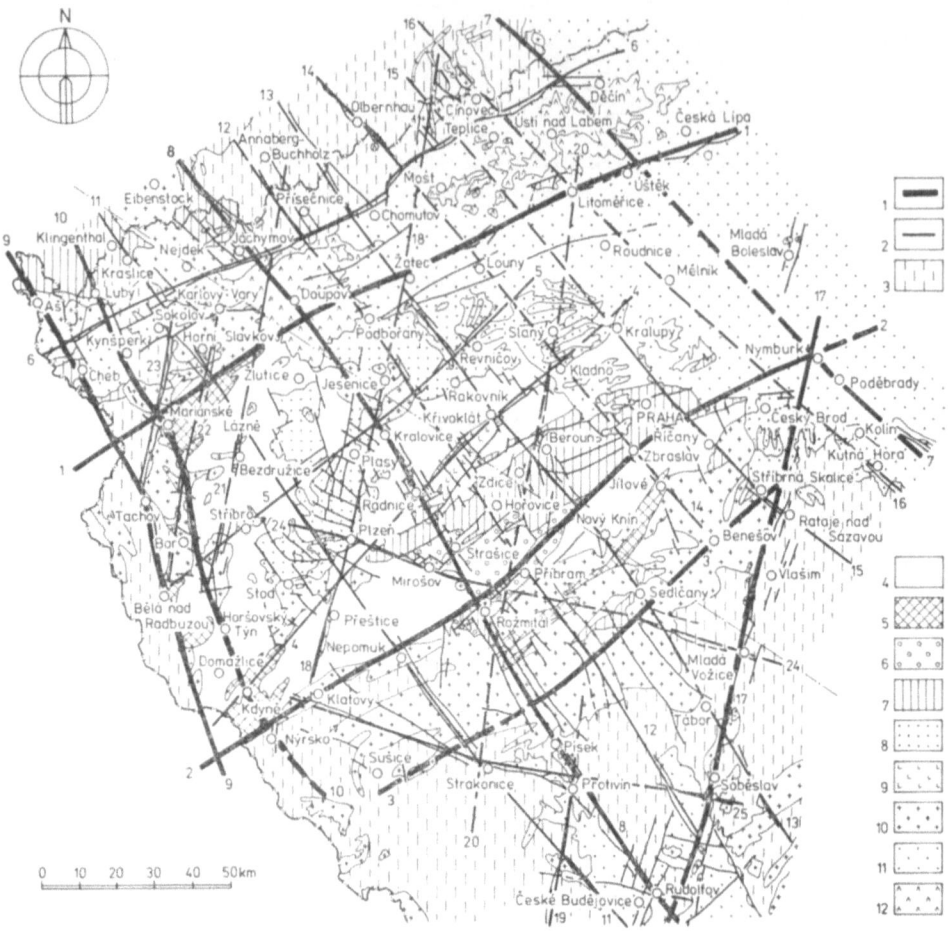

Fig. 1. Scheme of deep faults in the Central part of the Bohemian massif according to
Röhlich and Šťovíčková (1968)

1 — Main deep faults (1 Litoměřice, 2 Klatovy, 3 Benešov, 4 Kladno, 5 Stříbro, 6 Krušné
hory, 7 Poděbrady, 8 Jáchymov, 9 Tachov, 10 Mariánské Lazně, 11 Horní Slavkov,
12 Příbram, 13 Tábor, 14 Brandov, 15 Říčany, 16 Cínovec, 17 Blanice, 18 Plzeň, 19 Protivín,
20 Zdice, 21 Bezdružice, 22 Bor, 23 Kynžvart, 24 Mirošov, 25 Horažďovice); 3 — Moldanubian crystalline complex (in the Krušné hory Mts. block the crystalline series under
Ordovician); 4 — Upper Proterozoic; 5 — Mafic rocks of Upper Proterozoic age;
6 — Central Bohemian Cambrian; 7 — Ordovician up to Lower Carboniferous, 8 — Upper
Carboniferous and Permian; 9 — Palaeozoic porphyrites and porphyries; 10 — Granitoids;
11 — Upper Cretaceous and Tertiary sediments; 12 — Tertiary Volcanics

data of deep seismic sounding and then with the already known geological
faults on the surface and, finally, with volcanic, plutonic and metallogenic
features.

The individual fault systems are characterized primarily by their general directions which can substantially change due to deformations in the areas of higher tectonic mobility.

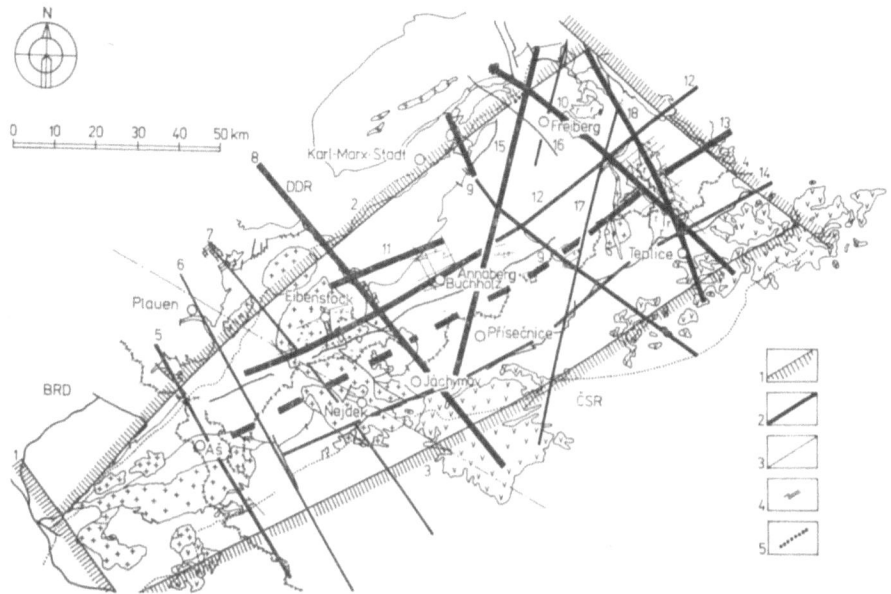

Fig. 2. Structural scheme of the Smrčiny and Krušné hory Mts. anticlinal zone after Hösel (1972)

1 — Lineament, 2 — Deep fault, 3 — Structural line, 4 — Veins of Sn-formation, 5 — Felsit horizont Numbers of respective faults: *1* Fränkisch line, *2* Central Saxonian lineament, *3* North Bohemian lineament, *4* Elbe lineament, *5* deep fault of Aš-Cheb-Tachov, *6* Mariánské lázně deep fault, *7* deep fault of Netschkau-Brunn-Černava, *8* Gera-Jáchymov deep fault, *9* Flöhatal deep fault, *10* deep fault of Niederbobritsch-Schellerhau-Krupka, *11* deep fault of Aue-Geyer, *12* Mittelerzgebirge deep fault, *13* Süderzgebirge deep fault, *14* boundary Erzgebirge deep fault, *15* deep fault of Nossen-Marienberg-Jöhstadt, *16* deep fault of Frieberg-Brand, *17* deep fault of Frauenstein-Seiffen, *18* deep fault of Meissen-Teplice

In the last decade also several geologic-gravimetrical profiles through the Bohemian Massif have been calculated and geologically interpreted. Two parts of one of these deep cross-sections roughly coinciding with the profile of deep seismic sounding No. 6 (Beránek et al., 1971) are also presented here (Figs. 5 and 6).

2. Main Fault Systems in the Bohemian Massif

2.1. The most continuous system of fault structures is the NE–SW system of the so-called "Variscan" structures within the Teplá-Barrandian block (Pokorný et al., 1970), i. e. between the Litoměřice deep fault (No. 1 in Figs. 1 and 2, A4 in Fig. 4), the Benešov deep fault (No. 3 in Figs. 1 and 2, No. A7 in Fig. 4). and the Poděbrady deep fault (No. 4 in Fig. 1, No. B7 in Fig. 4) The geophysical indications of these deep

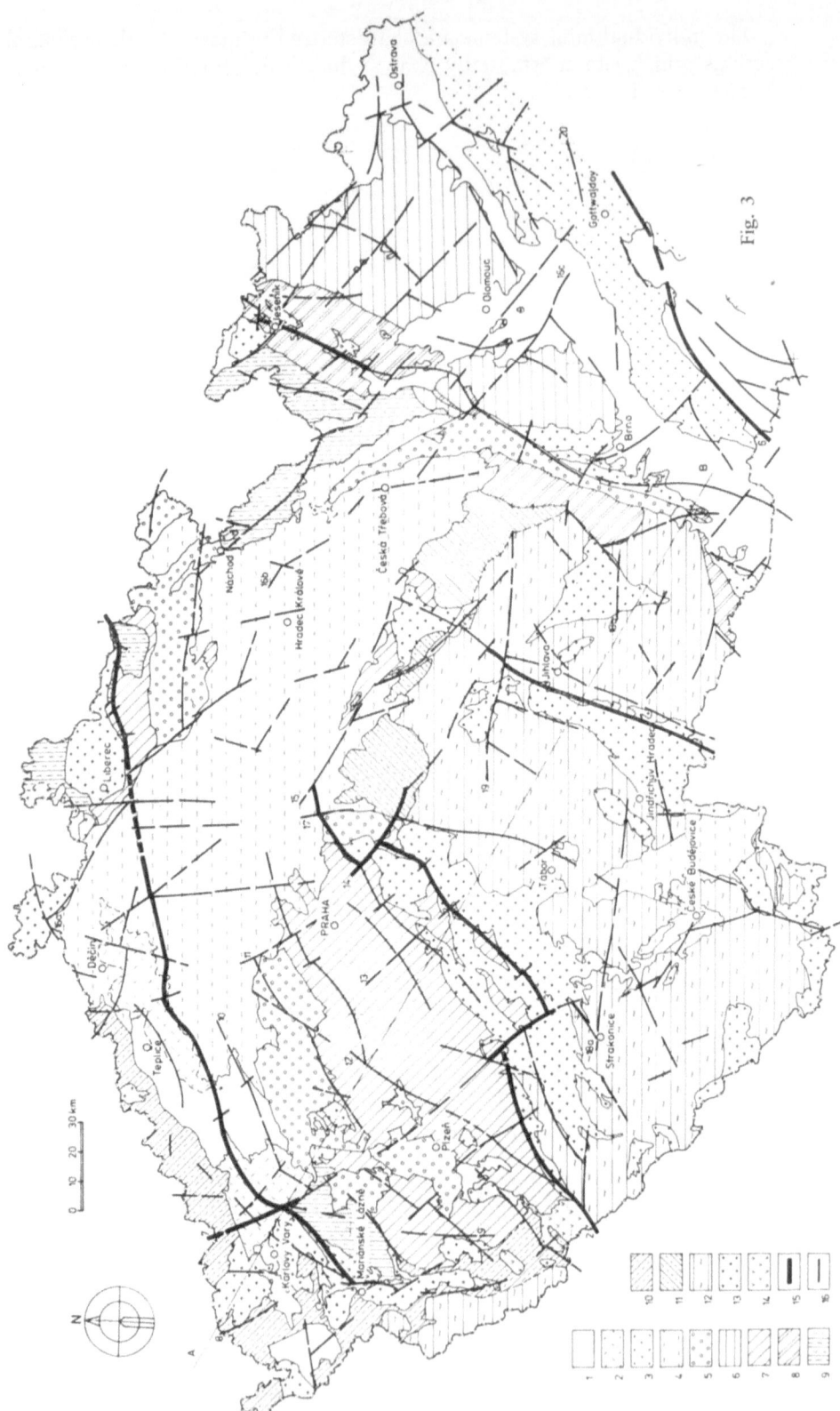

Fig. 3

faults can be followed practically through the whole studied region and are mostly evidenced on all depth levels. The mentioned boundary faults of the Teplá-Barrandian block are documented also by deep seismic sounding. This fault system forms a fundamental structural pattern of the block and divides it into a series of NE–SW trending gravity elevations and depressions correlating often with regional magnetic anomaly zones. The gravity and magnetic elevations correspond to belts of mafic paleovolcanics while the depressions reflect the zones of granitoid plutonism. The indicated discontinuities contour often the younger platform cover of postorogenetic basins and themselves are the feeding channels for nevolcanic complexes.

In the Moldanubian block (in the Czech branch) these structures are not indicated while in the Moravian block (Weiss, 1977) they are again well expressed. According to the mutual position of indications in different depth levels we can suggest the direction, sense and angle of dip of disjunctive planes. The most expressive example is the Ohře rift (Kopecký et al., 1970) where the southern boundary deep fault of Litoměřice is dipping to the NW while the Central fault of České Středohoří Mts. and the Krušné hory fault (No. 14 in Fig. 3) have a dip to SE (cf. Fig. 5). To the south, the segment between structures A4 and A5 (Fig. 4) represent a long horst structure evidenced by occurrences of the Krušné hory crystalline at shallow levels (Opárno and Maršovice crystalline islets), i. e. in the block which, from the aspect of deep structure, belongs to the Teplá-Barrandian block. Also, further structures of this system (A6 and A7 in Fig. 4) — Stříbro and Kladno faults in Fig. 1 dip to the SE — the dip of the Stříbro discontinuity being nearly vertical, which corresponds to the principle of diapiric penetration of the Čistá granodiorite stock connected with that fault (Bartošek et al., 1969). The Kladno fault is a feeding channel of the Křivoklát-Rokycany volcanic zone of Cambrian age. The Central Bohemian transitional segment (Röhlich, Šťovíčková, 1968; Pokorný et al., 1970) between the Klatovy and Benešov faults has probably a much more complicated deep structure than the above mentioned horst segment of Ohře. In the geological history, the character of this transitional segment has periodically changed, once being a horst, in other tectogenic phases a graben. The volcanic Jílové zone connected with an expressive magnetic anomaly is situated near the axis of this segment.

Fig. 3. Structural scheme of the Bohemian Massif based on geophysical indications according to Blížkovský, Pokorný, Weiss (1975)

1 — Neogene sediments; 2 — Neovolcanics; 3 — Flysch zones of the Carpathians; 4 — Cretaceous and Triassic rocks; 5 — Permo-Carboniferous sediments; 6 — Palaeozoic of Moravia; 7 — Early Palaeozoic, Preterozoic; 8 — Crystalline complexes of the Moravo-Silesian region; 9 — Crystalline complexes of the West Sudeten; 10 — Crystalline complex of the West bohemian region; 11 — Kutná Hora crystalline complex; 12 — Moldanubicum; 13 — Granitoid rocks; 14 — Basic and ultrabasic rocks; 15 — Deep fault zones; 16 — Major regional faults (1 Litoměřice deep fault, 2 Klatovy deep fault, 3 Benešov deep fault, 4 Přibyslav and Hlinsko zone, 5 Červenohorské sedlo zone, 6 Lednice zone, 7 Jáchymov fault, 8 Horní Slavkov line, 9 Bezdružice line, 10 Žlutice line, 11 Stříbro line 12 Kladno line, 13 Tábor line, 14 Sázava (Říčany) line, 15 Železné hory fault, 16a Lusitian fault, 16b Jílovice fault, 16c Nectava-Konice zone, 17 Blanice furrow, 18a Horaždovice line, 18b Třebíč fault, 19 Vír line, 20 Holešov zone

Generally, all structures of the NE−SW fault system represent the old tectonic lines rejuvenated in the Variscan tectogenesis. The Ohře rift, and in lesser degree also the Elbe mobile zone, were active also in Tertiary. The greatest accumulation of neovolcanic alkaline complexes of Doupov and České Středohoří Mts. are lying directly on the crossing of this rift with perpendicular deep fault structures, viz. the Jáchymov fault (B4 in Fig. 4) and the Poděbrady boundary fault of Elbe lineament (B7 in Fig. 4). These lengthwise NE−SW structures are often horizontally shifted by structures of another direction. According to either the horst or graben nature of the respective crustal segments, the faults of this system are the result of either compressive or tensional stresses in the earth's crust. They never have the character of strike-slip faults.

In the neighboring region, to the east of the Poděbrady fault (B7 in Fig. 4), i. e. within the Elbe mobile zone (lineament) this system appears in a somewhat different way than in the Teplá-Barrandian block. The structures are of lengthwise nature from the gravimetric and geological point of view, but they turn to a subequatorial direction and lose their dominant deep character (i. e. A4). This rotation of structures can probably be correlated with the recent results of palaeomagnetic studies (Krs, 1978).

2.2. The transverse system of structural discontinuities of NW−SE direction is less continuous than the above-mentioned one but the structures are much longer and continue through the blocks regardless of their boundaries. They are indicated on different depth levels, and often even the same structure has a different depth in respective intervals of its course. The vertical movements along them are local but the horizontal displacements are common. The latter are documented by frequent shifts and dislocations of lengthwise discontinuities along these faults. The dips are mostly vertical according to the Linsser-indications and thus it is not possible to suggest overthrust movements along them.

The most expressive structure of this system is the Jáchymov deep fault (Vondrová, 1963; Krs, Šťovíčková, 1966) — B4 in Fig. 4. It is remarkable that the transitional crustal segments (the Ohře horst segment and the Central Bohemian segment) have been developed only to the east of the Jáchymov fault, i. e. that the block boundaries have been shifted from the axis of the Teplá-Barrandian block outside. Moreover, Lower Palaeozoic volcanics occur within the sedimentary basin of Cambrian up to Devonian age (Barrandian s. s.) only in this probably dilated section of the Teplá-Barrandian block. The assumption of sea-floor spreading during that time span is reasonable and the interpretation of the Jáchymov deep fault as a paleotransform fault seems to be probable. The Jáchymov fault itself was the feeding channel of Cambrian porphyry and porphyrite linear volcanic subaeric eruptions causing one of the most expressive magnetic anomalies in the Bohemian Massif (Vondrová et al., 1964; Krsová, Šťovíčková, 1975).

Remarkable horizontal movements also took place along other structures of this system, i. e. along the Říčany (Sázava) fault and above all along the Poděbrady deep fault (B7 in Fig. 4), which is doubtlessly the western boundary fault of the Elbe mobile zone. In the segment above the Elbe lineament, we can trace the transverse system only with difficulty because the NW – SE faults are transverse only with regard to the deep crustal

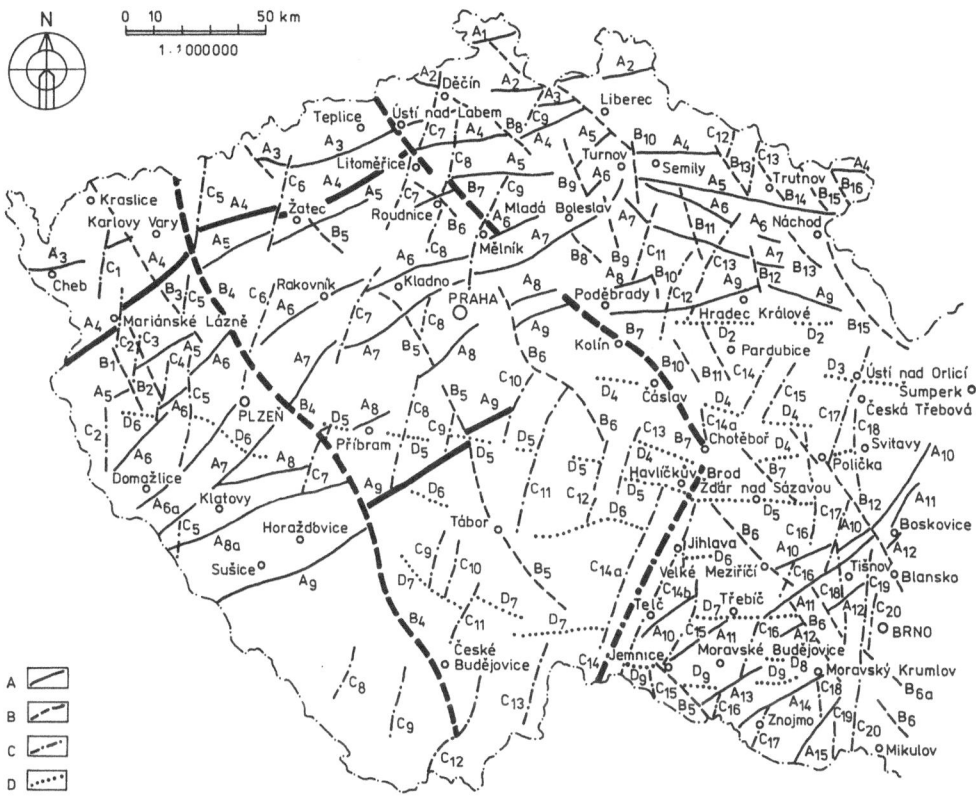

Fig. 4. Scheme of geophysical indications of fault system in the Bohemian Massif compiled by Pokorný and Šťovíčková

A — System of NE-SW faults; B — System of NW-SE (transcurrent) faults; C — System of submeridional (Rhine) faults; D — System of subequatorial faults

structure, but lengthwise with regard to the structure of the sedimentary cover which reflects the complicated tectonic development of the whole crustal segment. The tectonic pattern inside the Elbe mobile zone is generally more chaotic with signs of probable paleoorientation.

The NW – SE system of faults is well developed in the Moravian block. There, the idea of very old horizontal displacements along these transcurrent faults is suggestive judging from the internal structure of the Strážek Moldanubicum and the Svratka anticline.

The mostly clockwise but also anticlockwise horizontal movements along all SW – NE faults are surely of older age than alpine, as it has been suggested by Klomínský and Bernard (1974), probably with some phase

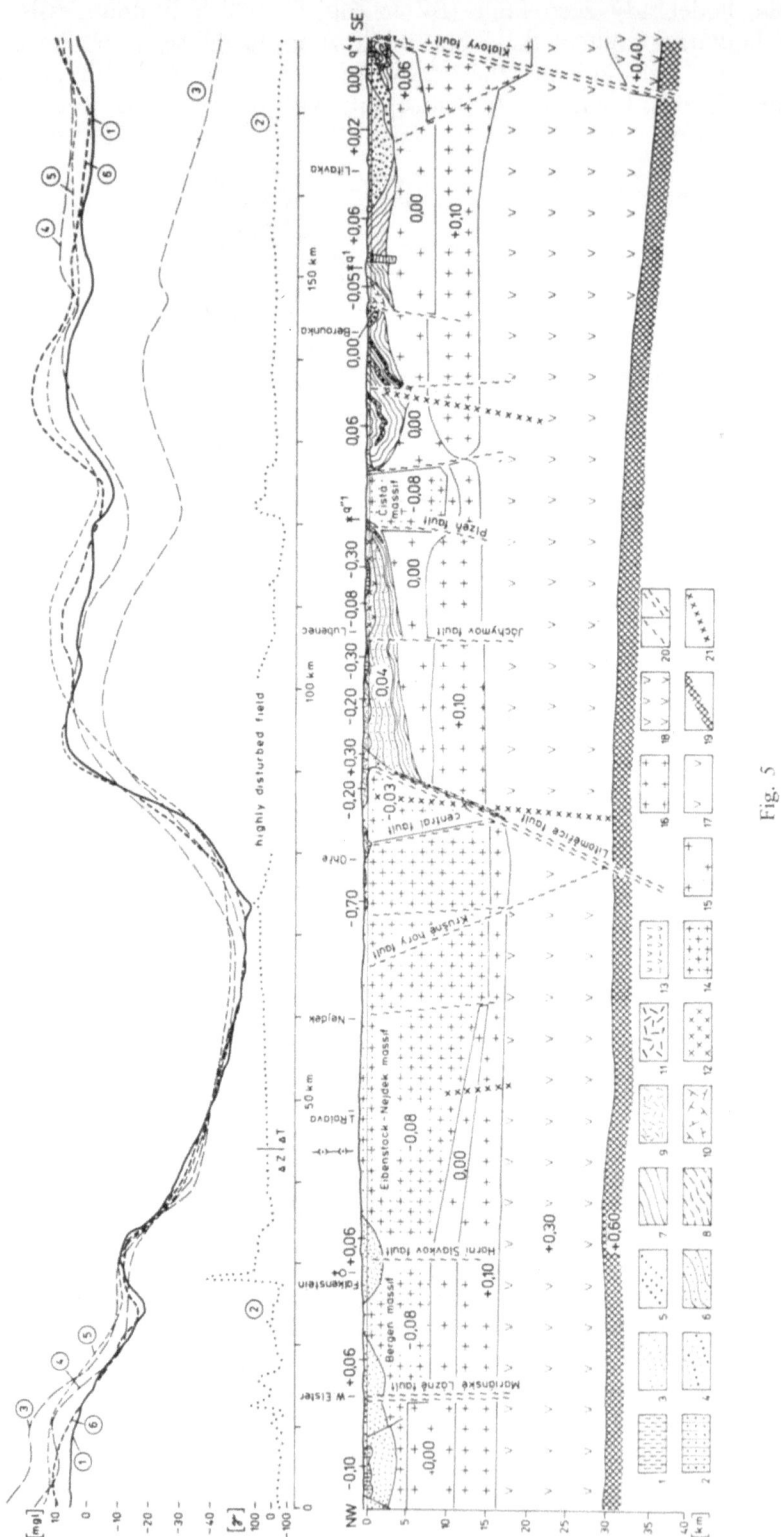

Fig. 5

of variscan tectogeny which dominantly influenced the distribution of crustal masses by huge plutonic processes and thus affected the gravity field in the Bohemian Massif.

This fault system is of prime importance for metallogenic processes since the tensional torsion ruptures favourable for hydrothermal ascension originate during great horizontal displacements of crustal blocks.

2.3. The system of submeridional faults of the Rhine (NNE–SSW) direction is also well developed in the Bohemian Massif, but its character is quite different from the above-described two fundamental diagonal systems. It is less connected with some definite phase of tectogenesis and has been operating evidently since the oldest geological epochs. The submeridional faults are well expressed in the younger Variscan age as well known long furrows filled either continuously or incoherently by Carboniferous and Permian sediments. The geophysical indications of these faults are mostly interrupted, but they are very stable in their azimuthal orientation even in the mobile zone of Elbe lineament and Moravian block. Similarly to the NW–SE faults, they cross the block boundaries. They have evidently the nature of long tensional or compressional ruptures, i. e. either of grabens or horsts. The Blanice and Boskovice furrows, the Plzeň deep fault (C6 in Fig. 4) belong to a graben system, the Přibyslav (Jihlava) zone (C14 in Fig. 4) is, according to geophysical indications and widespread plutonism (emplacement of the Central Moldanubian pluton), of horst-character. This deep fault of the first order, shown also by deep seismic sounding, divides the Moldanubicum into two independent crustal blocks (Pokorný et al., 1970) with different gravity characteristics (Buday, Dudek, Ibrmajer, 1969). In this sense it represents an exception within the Rhine fault system in the Bohemian Massif, being itself a block boundary of the Moravian block characterized by chaotic tectonic patterns similar to those of the Elbe zone (Pokorný et al., 1979). In contrast to the Přibyslav deep fault zone, the well known Boskovice furrow is markedly developed in the surface tectonics within the Moravian bliock; it is not geophysically continuously indicated and thus has no deep roots like other submeridional faults. This phenomenon is probably due to the mentioned greater mobility of the whole eastern boundary of the Bohemian Massif megablock revealing specifically

Fig. 5. Deep cross section through the northwestern part of the Bohemian Massif compiled by Polanský and Šťovíčková (in Polanský, 1973). Course of profile line is drawn on Figs. 1, 2 and 3

1 — Tertiary; 2 — Permo-Carboniferous; 3 — Lower Palaeozoic undistinguished; 4 — Ordovician; 5 — Cambrian; 6 — Upper Proterozoic (Postspilitic stage); 7 — Lower stage of Upper Proterozoic; 8 — Lower Proterozoic (Gföhl orthogneisses); 9 — Neovolcanics; 10 — Acid volcanics undistinguished; 11 — Basic volcanics undistinguished; 14 — Granites of density 2.6 gcm^{-3}; 15 — Granitic layer of density 2.7 gcm^{-3}; 16 — Granitic layer with density 2.8 gcm^{-3}; 17 — Basaltic layer of density 3.0 gcm^{-3}; 18 — Basaltic layer of density 3.1 gcm^{-3}; 19 — Upper mantle; 20 — Deep faults; 21 — Seismic indications of deep faults. Geophysical curves: 1 curve of total Bouguer anomaly, 2 geomagnetic curve, 3 theoretical gravity curve for the first approximative model respecting the relief of Moho-discontinuity, 4 theoretical gravity curve for the second model not respecting the relief of Moho-discontinuity, 5 theoretical gravity curve for the second model respecting Moho, 6 theoretical gravity curve for the third model respecting Moho

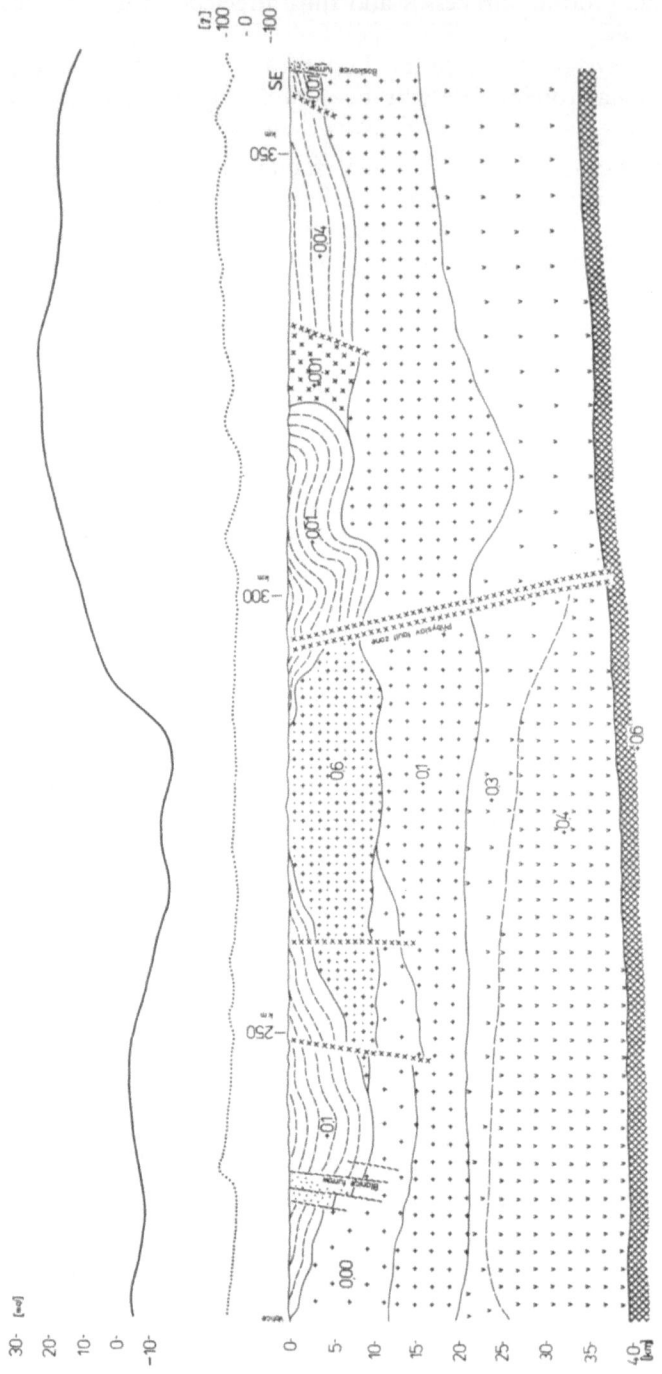

Fig. 6. Deep cross section through the central and southwestern part of the Bohemian Massif compiled by Mootlová and Weiss (in Polanský, 1973). This profile is not the direct continuation of Fig. 5. The section through the Central Bohemian transitional segment is omitted. Explanations are the same as in Fig. 5

the contamination of NNE−SSW and NNW−SSE directions. The Bosko-vice furrow is geologically defined as a synsedimentary furrow of multi-phase development with no signs of horizontal displacements (Jaroš and Mísař, 1967).

2.4. The subequatorial (WNW−ESE) system of faults, which is gener-ally perpendicular to the Rhine system, is mostly documented only in the Moldanubian block and in segments of directly adjoining blocks. The indi-cations are similar to those of submeridional system with the exception of evidenced horizontal movements. In no case do the subequatorial faults have the function of block boundaries; this is the reason why they have been underestimated in all tectonic concepts of the Bohemian Massif. How-ever, for recent ore exploration studies, they have great metallogenetic sig-nificance. The most expressive lines of this system are the Mirošov line (D6 in Fig. 4) and the Horaždovice line (D7 in the Fig. 4).

3. Geodynamic Concept of Tectonic Stresses in the Bohemian Massif

The recent tectonic stresses in Central Europe have recently been mea-sured by direct in-situ measurements (Greiner, Illies, 1977; Illies, Grei-ner, 1978) and then the results are suggested to support some geotectonic model in favor of plate tectonics concepts. The neotectonic stress field has been estimated from statistical values of orientations of joints or river val-leys (Scheidegger, 1978, 1979). Remote sensing methods can also be used for this purpose (Kronberg, 1975). Sometimes, a complex of data from different sources is used in order to evaluate the history of the stress field (Knetsch, 1967, 1969; Röhlich, Šťovíčková, 1968; Květ, Špička, 1973; Illies, 1978).

The comparison of surface fault systems with deep faults and lineaments has been studied (Šťovíčková, 1973; Metz, 1978) as well as the regularities of space orientation of fault patterns both on the Earth and terrestrial planets (Katterfeld, Charushin, 1973; Katterfeld, 1976; Šťovíčková, 1978). Some concepts of relation between geological processes and deep fault tectonics are based on fixistic ideas denying the role of substantial horizontal displacements along fault systems in the Bohemian Massif (Zeman, 1971, 1978) while another models suggest great horizontal motions (Klomínský, Bernard, 1974; Krs, 1979).

The author's concept (Šťovíčková, 1966, 1967, 1973) based on the geotectonic and planetological theory of rotational dynamics and applying this theory to the block structure of the Bohemian Massif is able to syn-thetize all mentioned aspects from the point of view of historical geodynamic development of the studied region. The block structure of the Bohemian Massif, firstly specified by Zátopek (1948) from the zones of increased seismic mobility is at present widely accepted by Czechoslovak geological community. Both the diagonal and orthogonal patterns of fault systems, particularly the fundamental deep faults, divide the basement of the Bohe-mian Massif into blocks with essentially different geological structure and development. The recent diagonal pattern delimits these blocks. According to palaeomagnetic data (Krs, 1969, 1978) it is possible to suppose that the NE−SW faults were oriented close to a paleomeridian during the Devonian to Carboniferous era and that Central Europe was located in an equatorial belt. According to the principle of the deformation of the Earth's ellipsoid, the intense variscan tectogenesis is easily understandable just in the area of Central Europe. Here, the recent NE−SW trending structures had, from the aspect of paleoorientation, a meridional course. They exhibit the character of either rifts and grabens (with mafic paleovolcanics) or horsts and anticlines (with granitoid plutons). This basic structure is preserved in the recent gravity field of the Bohemian Massif and it is possible to assume that the main paleostress field in the Variscan epoch was oriented in an E−W direction, i. e. along the paleoequator which corresponds to the principle of deformation of the Earth's ellipsoid. By the theory of rotational dynamics, the horizontal movements and shifts along recent NW−SE faults (Kölbel, 1954) can be readily explained as well. These faults had approximately equatorial orientation during the Variscan tectogenesis and horizontal movements balanced the differences in inertial moments of adjoining blocks. The Elbe mobile zone with its complicated and chaotic structural pattern is a good example. The recent NNE−SSW (Rhine) system of furrows in the Bohemian Massif is also consistent with the paleostress field of predominating E−W tensional stress during the Upper Carboniferous and Lower Permian according to paleomagnetic data. The scatter of these data is probably due to plasticity of the lower layers of the Earth's crust. This phenomenon correlates well with the concept of diapitric penetration of small circular granitoid intrusions along these submeridional faults (Bartošek et al., 1969) leading occasionally to the origin of cryptovolcanic ring structures in the Bohemian Massif (Gnojek, Šťovíčková, 1974).

The rotational-dynamics concept is not contradictory to the plate tectonic movements suggested for the Alpine tectogenetic cycle. The neoidic mobility and revival of the Ohře rift, Elbe lineament as well as the complicated internal structure of the Moravian block neighboring to the West-Carpathian system support such ideas.

Acknowledgements

The results presented here would not have been reached without several years of co-operation with colleagues from the Institute of Applied Geophysics namely Dr. L. Pokorný, Dr. L. Beneš (computer data) and many others. The author expresses her sincere thanks to all of them.

References

Bartošek, J., Chlupáčová, M., Šťovíčková, N.: Petrogenesis and Structural Position of Small Granitoid Intrusions from the Aspect of Petrophysical Data. Sbor. geol. Věd., UG 8, 37—68 (1969).

Beneš, L., Pokorný, L., Šťovíčková, N.: Geofyzikální charakteristika hlavních zlomových struktur západní části Českého masívu. Výzkum hlubinné geologické stavby Československa, 177—185 (1975).

Beránek, B., Dudek, A., Suk, M.: Geologická interpretace hlubinného seismického sondování v ČSSR. Geol. průzkum *13* (12), 353—357 (1971).

Blížkovský, M., Pokorný, L., Weiss, J.: Structural Scheme of the Bohemian Massif based on Geophysical Data. Věst. ústř. Úst. geol. *50* (1), 1—8 (1975).

Buday, T., Ibrmajer, J., Dudek, A.: Některé výsledky interpretace gravimetrické mapy ČSSR 1 : 500000. Sbor. geol. Věd, UG *8*, 7—35 (1969).

Hösel, G.: Fortschritte der Metalogenie im Erzgebirge, Geologie *21* (4/5), 437—456 (1972).

Gnojek, I., Šťovíčková, N.: The Ring Structure of the Sedmihoří Granite Stock. Sbor. geol. Věd, UG *12*, 113—130 (1974).

Greiner, G., Illies, J. H.: Central Europe: Active Residual Tectonic Stresses. Pageoph. *115*, 11—26 (1977).

Illies, H.: Neotektonik, geothermale Anomalie und Seismizität im Vorfeld der Alpen. Oberrhein. Geol. Abh. *27*, 11—31 (1978).

Illies, J. H., Greiner, G.: Rhinegraben and the Alpine System. Bull. Geol. Soc. Am. *89*, 770—782 (1978).

Jaroš, J., Mísař, Z.: Problém hlubinného zlomu boskovické brázdy. Sbor. geol. Věd, G *12*, 131—147 (1967).

Katterfeld, G. N., Charushin, G. V.: General Grid System of Planets. Modern Geology *4*, 253—287 (1973).

Katterfeld, G. N.: Global and Regional Sytsems of Lineaments on the Earth, Mars and the Moon. Proceed. of the 1st Intern. Conf. on the New Basement Tectonics. Utah Geol. Ass. Publ. *5* (1976).

Klomínský, J., Bernard, J. H.: Segmentation of the Bohemian Massif in the Light of Variscan Magmatism and Metalogeny. Věst. ústř. Úst. geol. *49* (3), 149—159 (1974).

Knetsch, G.: Changing Tectonic Roles of the Upper Rhine Lineament in the Course of Geological Times and Events. Rhinegraben Progress Report (1967).

Knetsch, G.: Über Funktionswechsel des Rheinischen Lineamentes und die Entstehung des Oberrhein-Grabens. Zeitschr. Deutsch. Geol. Ges. *118*, 222—235 (1969).

Kölbel, H.: Große Seitenverschiebungen und Horizontalflexuren im Deutschen Grundgebirge und ihre lagerstättenkundliche Bedeutung. Geologie *3*, 445—450 (1954).

Kopecký, L., Dobeš, M., Fiala, J., Šťovíčková, N.: Fenites of the Bohemian Massif and the Relations Between Fenitization, Alkaline Volcanism and Deep Fault Tectonics. Sbor. geol. Věd, G *16*, 51—112 (1970).

Kronberg, H.: ERTS Data on Regional Fracture Patterns of Central Europe. Geodynamics Project, National Committee of the FRG (1975).

Krs, M.: The Scope of Palaeomagnetism in Geology. Sbor. geol. Věd, UG *7*, 43—80 (1968).

Krs, M.: Palaeomagnetic Evidence of Tectonic Deformation of Blocks in the Bohemian Massif. Sbor. geol. Věd, G *31*, 141—150 (1978).

Krs, M., Šťovíčková, N.: Palaeomagnetic Investigation of Hydrothermal Deposits in the Jáchymov (Joachimsthal) Region, Western Bohemia. Appl. Earth Sci. Trans./Sect. B of the Inst. of Min. and Metal. *75*, 51—57 (1966).

Krsová, M., Šťovíčková, N.: Petrofyzikální a petrologické problémy tektonického uzlu v oblasti rožmitálské osni rampy. Výzkum hlubinné geologické stavby Československa, 163—176 (1975).

Květ, R., Špička, V.: O genezi zlomů ve vztahu k vývoji sítě a systému puklinových zón a poruch v širší oblasti Vídeňské pánve. Geol. práce, Správy 60, 237—257 (1973).

Metz, K.: Bruchsysteme und Westbewegungen in den östlichen Zentralalpen. Mitt. Österr. Geol. Ges. 69, 27—47 (1978).

Palivcová, M., Šťovíčková, N.: Volcanism and Plutonism of the Bohemian Massif from the Aspect of its Segmented Structure. Krystalinikum 6, 169—199 (1968).

Polanský, J.: Hloubkové řezy Českým masívem. Geol. průzkum, 15, 6, 161—167 (1973).

Pokorný, L., Beneš, L., Šťovíčková, N.: Geofyzikální projevy strukturně geologických deformací. MS Geofond Praha — Vol. I. 57 p. (1975), Vol. II. 46 p. (1975).

Pokorný, L., Beneš, L., Čuta, J., Šťovíčková, N.: Geofyzikální projevy strukturně geologických deformací. MS Geofond Praha, Vol. 3, 41 p. (1978), Vol. 4, 40 p. (1979).

Pokorný, L., Polanský, J., Šťovíčková, N.: Evidence of Deep Segmented Structure of the Bohemian Massif Based on Geophysical Data. Sbor. geol. Věd, UG 9, 7—18 (1970).

Röhlich, P., Šťovíčková, N.: Die Tiefenstörungs-Tektonik und deren Entwicklung im Zentralen Teil der Böhmischen Masse. Geologie 17 (6/7), 670—694 (1968).

Scheidegger, A. E.: Joints in the Southern Alps. Geol. appl. e idrogeol. 13, 129—139 (1978).

Scheidegger, E. A.: Orientationsstruktur der Talanlagen in der Schweiz. Geograph. Helvetica 34 (1), 2—8 (1979).

Šťovíčková, N.: Eine Theorie der Erdevolution auf der Basis von Rotationsdynamik. Geologie 15 (10), 1123—1134 (1966).

Šťovíčková, N.: Genetické principy hlubinné zlomové tektoniky. Čas. min. geol. 12 (1), 69—78 (1967).

Šťovíčková, N.: Hlubinná zlomová tektonika a její vztah k endogenním geologickým procesům. (Deep Fault Tectonics and its Relation to Endogenous Geological Processes). 198 p. Academia Praha (1973).

Šťovíčková, N.: Comparative Tectonics of Terrestrial Planetary Bodies. COSPAR 21st Plenary Meeting, Program/Abstracts, 264—265 (1978).

Vondrová, N., Chlupáčová, M., Šalanský, K.: Výzkum vulkanismu v oblasti rožmitálského kambria (Erforschung des Vulkanismus im Kambriumgebiet von Rožmitál). Sbor. geol. Věd, G 5, 59—93 (1964).

Vondrová, N.: Hlubinné tektonické zóny v Českém masívu a jejich význam pro metalogenezi. Geol. průzkum 5 (6), 161—164 (1963).

Weiss, J.: Fundament moravského bloku ve stavbě evropské platformy. Folia fac. sci. nat. Univ. Purk. Brun. 18, Geol. 30, 5—65 (1977).

Zátopek, A.: Šíření východoalpských zemětřesení Českým masivem. Publ. Čs. stát. geof. úst., spec. práce 3 (1948).

Zeman, J.: Některé aspekty genetické klasifikace hlubinných zlomů v Českém masívu. Geol. práce, Správy 57, 175—188 (1971).

Zeman, J.: Deep-Seated Fault Structures in the Bohemian Massif. Sbor. geol. Věd, G 31, 155—185 (1978).

Address of author: Dr. N. Šťovíčková, Inst. Applied Geophysics, Podbělohorská Silnice 47, CS-15000 Praha, Czechoslovakia.

Rock Mechanics, Suppl. 9, 139—146 (1980)

Rock Mechanics
Felsmechanik
Mécanique des Roches
© by Springer-Verlag 1980

Mathematical Representation of the Fracture Field in the Venetian Alps (NE Italy). Preliminary Results

By

F. Degan, E. Martino, F. Pianetti, and A. Zanferrari

With 10 Figures

Abstract

We propose a simple model for understanding the apparently disordered pattern of lineaments observed on a LANDSAT-1 Frame of the Venetian Alps. The lines have been tested on the field for geological meaning. The number of lines observed is 4111. We have been able to recognize four principal modes, with a symmetrical pattern. The width of the distribution (σ) can be (a little) reduced by supposing a variation of the azimuth with the coordinates x and y. The resulting field of directions is represented in a suggestive way by integrating the differential equation implied in the formula $\vartheta = f(x, y)$. A deeper statistical test suggests that the pattern of four modes should be reduced to just two modes. We are carrying out more computations to test the significance of the model proposed with better resolution.

The region with which we are concerned is a part of the Italian Alps to the North of Venice. Our aim is to recognize in the fracture field the most frequent and expressive directions in order to reconstruct, at a later time, the tectonic stress field. The region is subject to a non-negligible seismic activity. Fault plane solutions have been obtained for many recent earthquakes.

It seems interesting to compare such solutions with the results of the analysis of the fracture field. The region extends between the Adige and Piave rivers. From a geological point of view the area belongs to the Southern Alps. Outcropping rocks belong mostly to the post-Variscan cover (Permian-Tertiary, with modest Quaternary deposits); rocks of the basement are locally present, mostly metapelites and granitoids.

The fracture field has been reconstructed from Landsat-1 frame *n*. 1218—09335 band 7 (near infrared), *b* and *w* positive print, scale 1 : 250 000 (Fig. 1). On this frame we have at first plotted linears of length 0.3 to 6 cm, corresponding, on the field, to lines of a length of 0.75 to 15 km

The tectonic nature of such lines is based on previous direct knowledge and on *ad hoc* survey of test areas: lines correspond mostly to faults of various importance and meaning. Lineaments are plotted out in Fig. 2;

0080-3375/80/Suppl. 9/0139/$ 01.60

Fig. 1 is the Landsat frame. Fig. 3 represents the density distribution of lines along azimuth. Fig. 4 represents distribution of lines weighted with their length. Both figures, 3 and 4, suggest that some directions are more densely provided with lineaments.

The information is represented in another way on Fig. 5, which represents lineaments comprised in intervals between fixed values of the azimuth.

Fig. 1. LANDSAT-1 Frame n. 1218-09335/7 — Feb. 26, 1973

Fig. 5 also shows that lineaments in each interval, although with various densities, always fill all the field and do not appear concentrated on particular locations.

For exploiting the features of the plots, we have tried to understand the pattern of lineaments as a superposition of various fields of directions, each of which is expressed by a suitable polynomial in x and y.

The first approximation of each field can be expressed by a constant function. Subsequent refinements, which introduce at first linear terms, then terms of second or higher order, are carried out with the aim of reducing the dispersion of the residuals, i. e. of the differences at each point between the computed value and the observed one.

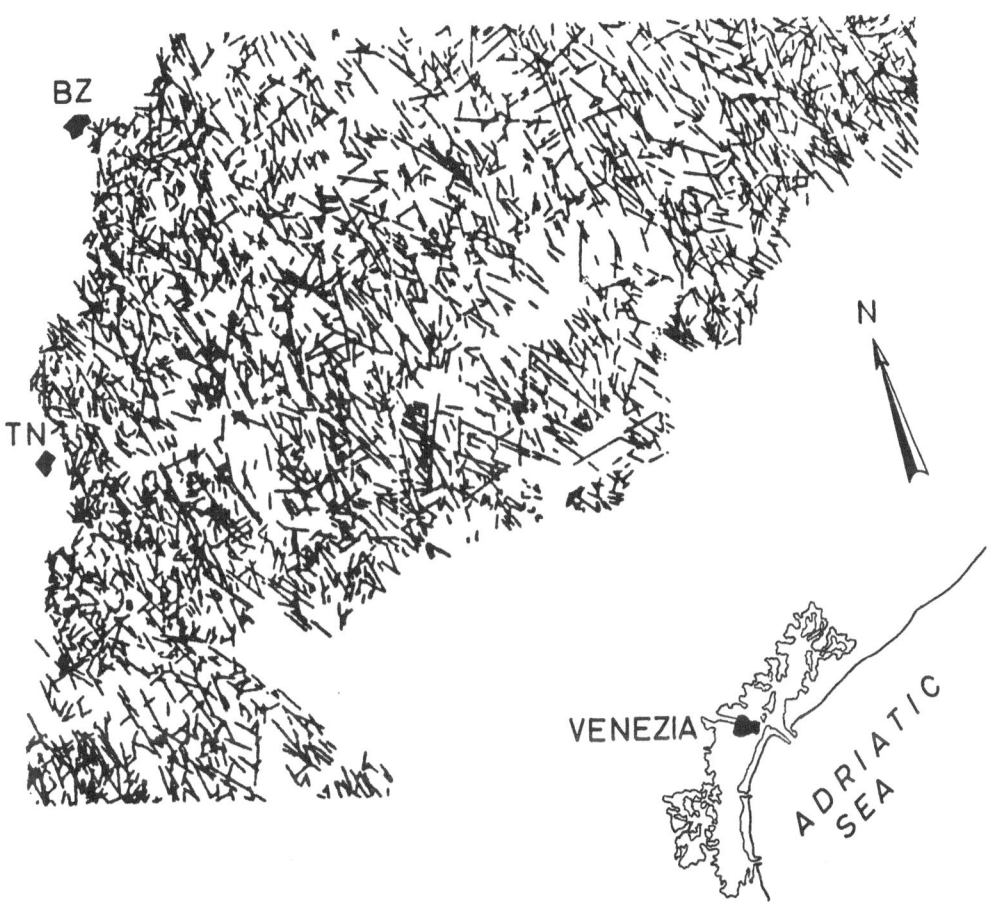

Fig. 2. Observed lineaments total field (TN: Trento; BZ: Bolzano)

A part of such residuals can never be eliminated, due to the natural dispersion of any physical measurement, but another part of the residuals may be due to spatial variation of the preferred direction of lineaments.

To display the spatial variation of the direction of lineaments a first investigation was performed by a "best fit" technique. Such an investigation results in a "most likely direction" for each point of the field. Subsequently, only those directions that differ from the "most likely" one by less than a suitable tolerance deduced from the graph (Fig. 3), are taken into consideration. This kind of filtering allows one to find higher order polynomial ex-

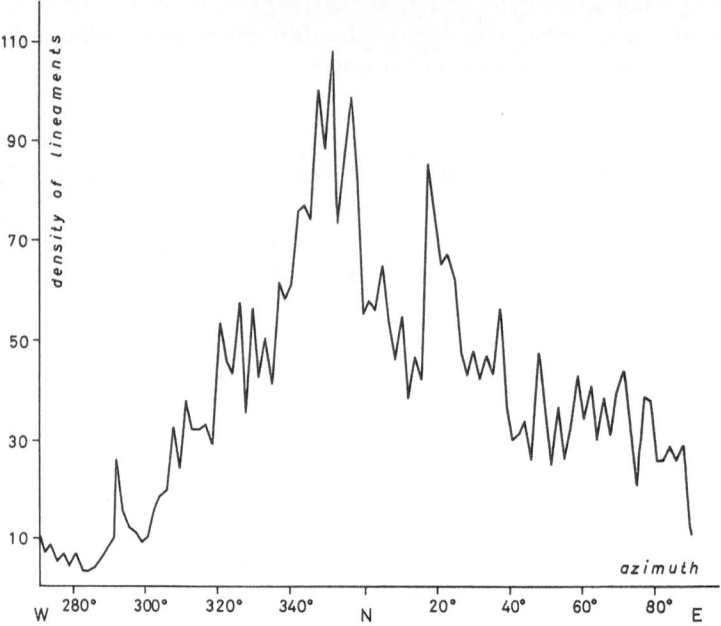

Fig. 3. Number of lines per azimuth interval (400 intervals in π)

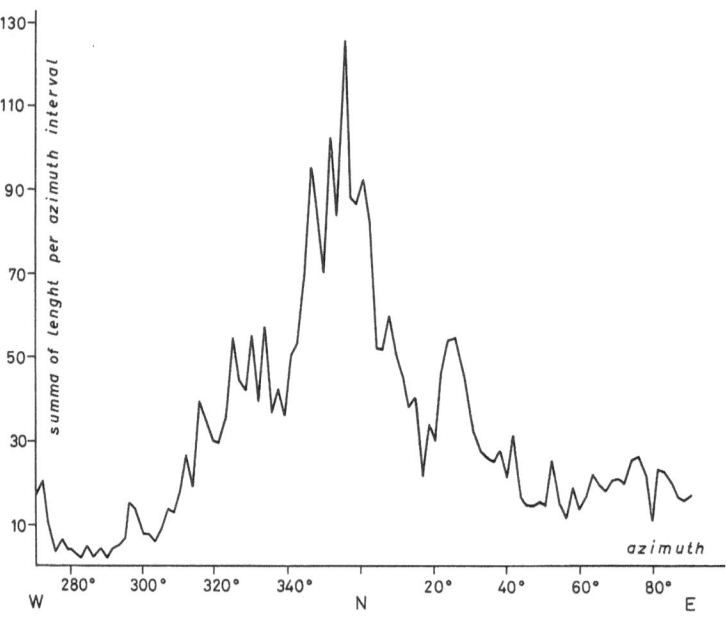

Fig. 4. Sum of length of lines in the intervals of Fig. 3

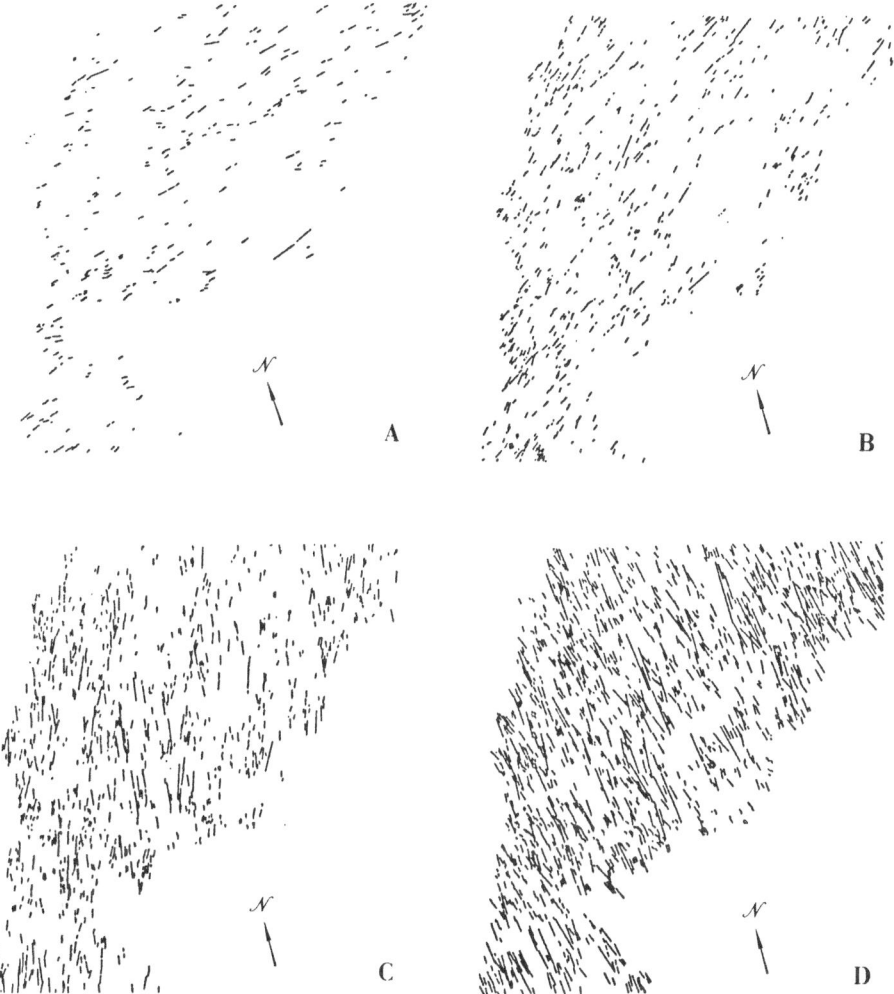

Fig. 5. Observed lineaments with azimuth between:

N 80° E — N 115° E (A)
N 45° E — N 80° E (B)
N 10° E — N 45° E (C)
N 25° W — N 10° E (D)

pressions for ϑ. Such expressions are used to formulate a pair of differential equations

$$dx/ds = \cos \vartheta$$

$$dy/ds = \sin \vartheta$$

This set is then integrated producing the family of curves shown in Fig. 6.

It is necessary to ask if the dispersion of the experimental data is really reduced by introducing linear and higher order terms in the description of the field. We have to say that present computations still leave a margin of doubt in such a reduction, so that it is necessary to perform a sufficiently powerful and discriminating numerical test. In addition, we have used the described filtering technique in order to separate various contributions suggested from the experimental data distribution.

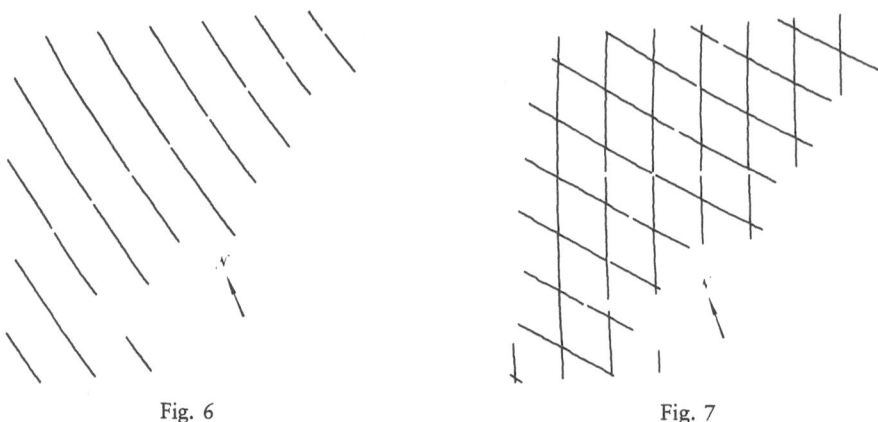

Fig. 6   Fig. 7

Fig. 6. Most likely direction (circa N 10⁰ W)

Fig. 7. Most likely directions near to N 20⁰ E and N 40⁰ W

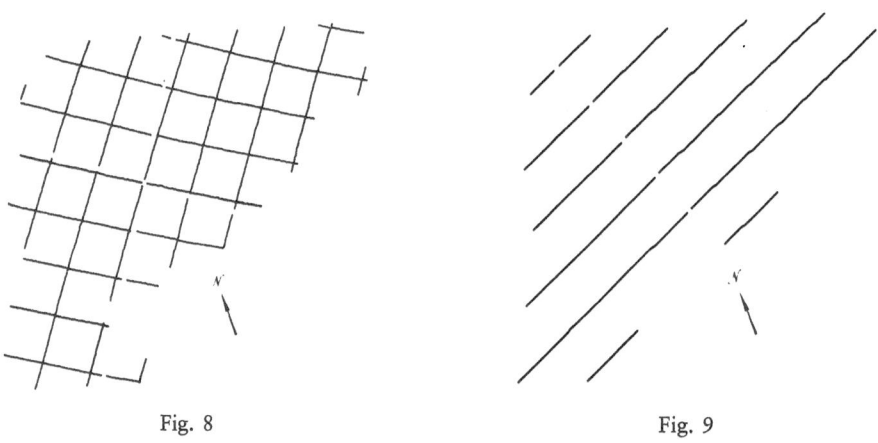

Fig. 8   Fig. 9

Fig. 8. Most likely directions near to N 45⁰ E and N 65⁰ W

Fig. 9. Residual directions (circa N 69⁰ E)

In the first test, we separated four components. We reconstructed the parameters of hypothetical Gaussian components by a logarithmic transformation, followed by the search for the best parabola. Parameters thus obtained were later used as a first guess for a complete fit. The complexity of the graph dissuades one from employing totally automatic methods and rather suggests an "operator controlled search", with later use of more precise tests for the final verification of results. Fig. 10 shows two such fitted parabolas.

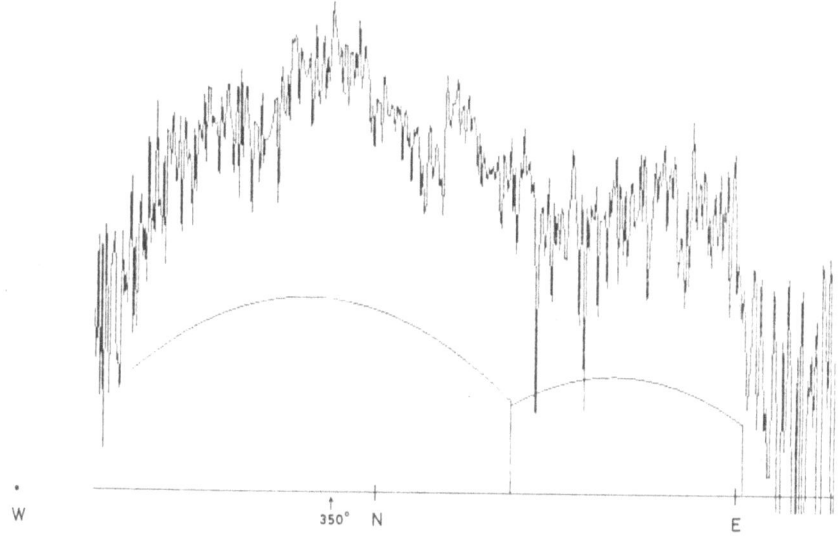

Fig. 10. Example of first search for the sigma parameter of the component distributions

Moreover, our technique has indicated a possibly symmetric pattern of two lower peaks around the greater one. Perhaps also the cluster between azimuth N 40⁰ and N 95⁰ has to be considered as composed of parts symmetrically posed around the greater one. It seems in fact that such cluster has a mean value at the angular distance of $\pi/2$ from the most enhanced peak. Finally, this procedure has shown the patterns shown in Figs. 6, 7, 8, 9.

To verify the utility of this decomposition, we have used a test for normality; it should be noted that this kind of statistical analysis is not a classical one. We are trying to separate the contributions of various clusters which are partially mixed and we want to test their normality. A first criterion could be to assign the experimental datum to the cluster with the nearest center. This distribution, although univocally definable, cannot later be tested for normality of its components: the tails, in fact, would be inexorably truncated. We had to resort to a "censored data theory" (we thank Prof. F. Pesarin of the Institute of Statistics of the University of Padua, who has kindly let us know some of his results still in press). The tails of the distributions are reconstructed by assigning each datum to a cluster or to another one by a stochastic sorting which refers to the amplitudes of each component of the distribution. Such sorting obviously reduces the resolution of the test, but the number of data should allow one to get enough resolving power. The work in this direction is in progress, but it is still incomplete.

Some of the recent results seem to suggest that two lower symmetrical peaks around the most enhanced one are not significantly distinguishable from the last one, so that only two directions are effectively present. We are also studying this problem.

Address of authors: F. Degan, Istituto di Matematica Applicata, Università di Padova, via Belzoni 7, I-35100 Padova, Italy; E. Martino, Gruppo Nazionale Informatica Matematica, Università di Padova, via Belzoni 7, I-35100 Padova, Italy; F. Pianetti, Laboratorio per lo Studio della Dinamica delle Grandi Masse, Consiglio Nazionale per le Ricerche, S. Polo 1364, I-30100 Venezia, Italy; A. Zanferrari, Istituto di Geologia, Università di Padova, via Giotto 1, I-35100 Padova, Italy.

Rock Mechanics, Suppl. 9, 147—158 (1980)

Rock Mechanics
Felsmechanik
Mécanique des Roches
© by Springer-Verlag 1980

Theme 4

Petrofabrics and Stresses

Palaeo- and Recent Stress Fields in Tunisia and Libya from the Cenozoic Structural Bearing

By

Karlheinz Schäfer

With 5 Figures

Abstract

Small-scale structural features have been applied as evidence for the existence of horizontal palaeo-stresses within Mesozoic and Cenozoic carbonates of Tunisia and northern and central Libya. Major tectonic elements such as the late Mesozoic evolution of the Sirte graben and horst structures, the Tunisian Atlas folds and grabens, and the Cenozoic Tibesti-Garian volcanic chain have also been considered in this analysis. It is shown that Tunisia and northwestern Libya were subjected to horizontal crustal shortening in the NE – SW direction from the end of the Cretaceous to the Upper Oligocene with a climax during the Middle to Upper Eocene. A different tectonic regime with mainly extensional movements to the NE and volcanic activity was contemporaneously effective from the Tibesti-Garian volcanic chain to the east. Northwestern Libya is presently influenced by a NW – SE-directed horizontal principal stress of a stress field that already existed during the Middle Miocene. This stress field had a climax during the Upper Pliocene inducing graben formation and folding in Tunisia, and also causing no major rifting in the Sirte basin.

It is concluded that the African plate while drifting NE from the Upper Cretaceous through the end of the Palaeogene and moving NW relative to stable Europe during the last 10 m. y. has induced the two tectonic regimes in Tunisia and in northern Libya.

The different crustal stress history in the Sirte basin may be due to a continuous NE motion of the African plate after the indentation of Europe during the Upper Cretaceous by the Adriatic promontory as the continuation of Africa from NW-Libya and Tunisia to the north.

1. Introduction

The geodynamic evolution of the Mediterranean during the Mesozoic and Cenozoic is controlled by the relative movements of the two framing plates of Africa and Europe. Since the European lithospheric plate is con-

10*

0080-3375/80/Suppl. 9/0147/$ 02.40

sidered to have been stable during that period, the African plate on its approach to Europe must have caused all the geological features of the interplate area.

But some authors also believe that Africa was stationary during the last 30 m. y. (Burke and Wilson, 1972). However, there is plenty of evidence for a considerable NW—SE approach of Africa and Europe during the past 10 m. y. Pitman and Talwani (1972) have shown the post-Palaeozoic evolution of the North Atlantic in many stages applying palaeomagnetic and deep-sea drilling results to their reconstruction. The relative movement of Africa deriving from this has since been used for numerous interpretations of the kinematics of the Mediterranean. During the period concerned in this paper the African plate should have drifted NW from the Middle Upper Cretaceous to 65 m. y. ago, SW from 65 m. y. ago to the time of the turn Palaeocene/Eocene, and N from then to the Upper Miocene (Channell and Horvath, 1976).

It is difficult to rely on any of those relative drift episodes to explain the geological processes occurring in, at least, the western Mediterranean. Pitman and Talwani (1972) have deduced the relative movements of Africa and Europe assuming a stationary North American plate. But North America was neither stationary nor did it perform an equivalent motion in respect of Europe and Africa. The North American plate moved NW from 180—80 m. y. ago (Nevada-Sevier-Columbia orogeny) while it was still connected with Europe via Greenland, and drifted SW from 80—40 m. y. ago (Laramide orogeny) while Davis Strait and Baffin Bay opened between Canada and Greenland and the East Pacific rift-system was overridden (Coney, 1971).

Sclater et al. (1977) have shown 12 stages of the Atlantic evolution, the last 5 of which (at 65, 53, 36, 21, and 10 m. y.) permitted the African plate only a minor sinistral rotation relative to Europe. This convergent motion of the two plates is insufficient considering the amount of compressive deformation in the Mediterranean during that period. The Alps were shortened by at least 300 km during the "meso-alpine" main deformation (Trümpy, 1973). The North African crust in the Atlas Mountains has been shortened by 50% (Bernoulli and Jenkins, 1974).

There are also extensional structures within the Mediterranean orogenic belt and the stressed European and African forelands. They are sensible indicators of the orientation and timing of principal stress components of large crustal stress fields. One of those stress fields with NE—SW-oriented maximum horizontal stress components ($\sigma_{H\max}$) has initiated the formation of the Central and West European grabens, also the Valencia-Genova rift that widened to the Ligurian Sea rotating the Corso-Sardinian block counterclockwise. Another major stress field with NW—SE-directed $\sigma_{H\max}$ succeeded the previous one during the Neogene. It was effective in Europe and in North Africa creating contemporaneously numerous structural features such as folds in the Western Alps, NE Spain, Betic Cordillera, and in the Maghreb countries, and grabens in NW Germany, central Spain, and in Tunisia.

As for northern Libya the concept of the two Cenozoic stress fields causing the tectonism of Europe and North Africa does not seem to be valid. The evolution of the Sirte basin with its graben and horst structures obviously contrasts with the two-stress-field-theory. During the Upper Cretaceous and Palaeogene when a maximum compressive stress was NE−SW-oriented elsewhere, a tensile stress in that direction must have caused the subsidence of the Sirte grabens, and when the Neogene NW−SE-striking maximum horizontal compression created folds and grabens in nearby Tunisia, the evolution of the Sirte grabens ceased.

In this paper it is attempted to unravel this contradiction by means of structural investigations of small-scale tectonic features such as horizontal stylolites, vertical veins or dikes and pop-ups. All these structures are unequivocal indicators of the orientation and the type of stress.

Horizontal (or transverse, or tectonic) stylolites (H-stylolites) are solution features like the vertical stylolites (V-stylolites). Both have formed under stress. The V-stylolites derive from solution of rock material due to pressure of the overburden, whereas H-stylolites are caused by horizontal compressive stress. Stylolites form readily in pure limestones if there is sufficient stress and a solvent present. With increasing contamination or dolomitization of the limestones stylolitization decreases.

2. Results and Interpretation

In Tunisia horizontal stylolites were found and investigated at 29 sites. They were formed within limestones of Liassic to Burdigalian age. This was the first discovery of horizontal stylolites in Cenozoic sediments. Two distinct generations of H-stylolites are developed, an earlier set of H-stylolites in Palaeogene and Mesozoic sediments which is NE−SW-oriented, and a younger series of H-stylolites in Miocene and older limestones that is NW−SE-directed. Fig. 1 shows the sites and the orientations of the older generation of H-stylolites in Tunisia. It is suggested that these horizontal stylolites were formed mainly during the Upper Eocene. Investigations of H-stylolites in Spain (Schäfer, 1978) revealed that there is consistency in the age and in the orientation of the horizontal stylolites which may have formed under the influence of a large stress field with a $\sigma_{H\max}$-orientation of NE−SW. The orientation of the younger post-Burdigalian generation of horizontal stylolites in Tunisia is shown in Fig. 2. They were formed under the influence of a younger regional stress field whose $\sigma_{H\max}$ direction was NW−SE. It is suggested that the younger stress regime began after the Burdigalian and is still active at present according to *in situ* strain data in NW-Libya (Schäfer, 1980, this volume).

The area of study in Libya is underlain by Mesozoic and Cenozoic rocks of which carbonates are predominant. From the Azizia Formation of the Middle Triassic to the Gargaresh Formation of the Quaternary there is no major unit devoid of solid, more or less pure limestones that meet the lithological conditions of stylolitization. H-stylolites indicate the amount of horizontal compaction. Carbonate-filled veins may form contemporaneously striking in the direction of $\sigma_{H\max}$ and opening in the direction of least compressive or tensile stress ($\sigma_{H\min}$). They are commonly filled with calcite that derive from the solution of limestones during stylolitization. Calcite-filled veins indicate the direction of minimum principal stress and the amount of rock dilatation.

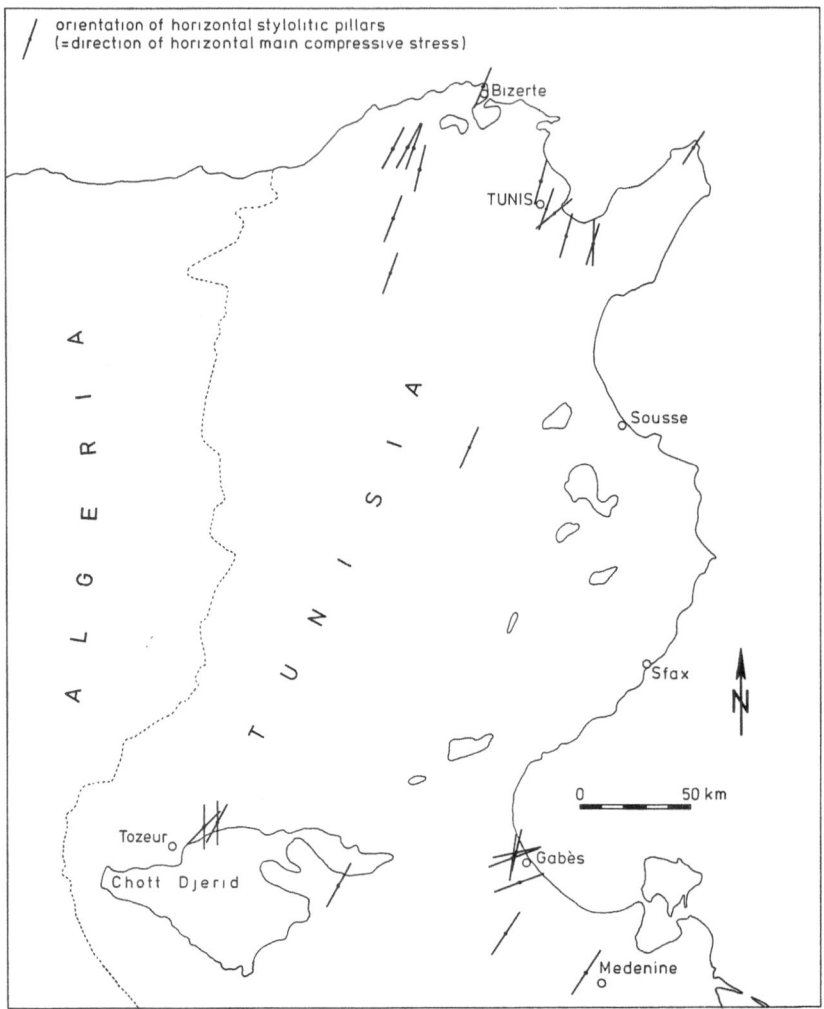

Fig. 1. NE-SW-oriented horizontal stylolites were formed from the Upper Cretaceous to the end of the Oligocene. It is the older set of pressure-solution features which indicates the $\sigma_{H\max}$-direction of a stress field resulting from the contemporary NE-SW-approach of Africa and Europe

Horizontal stylolites and calcite-filled veins have been observed in Tripolitania, on the flanks of the southern Hon graben and in Cyrenaica. They are developed in the Triassic Azizia Formation, Jurassic Bu Gheilan

Formation, Cretaceous Ain Tobi, Jefren, and Garian Formations, and Palaeogene Appollonia, Derna, and Jabal Waddan Formations.

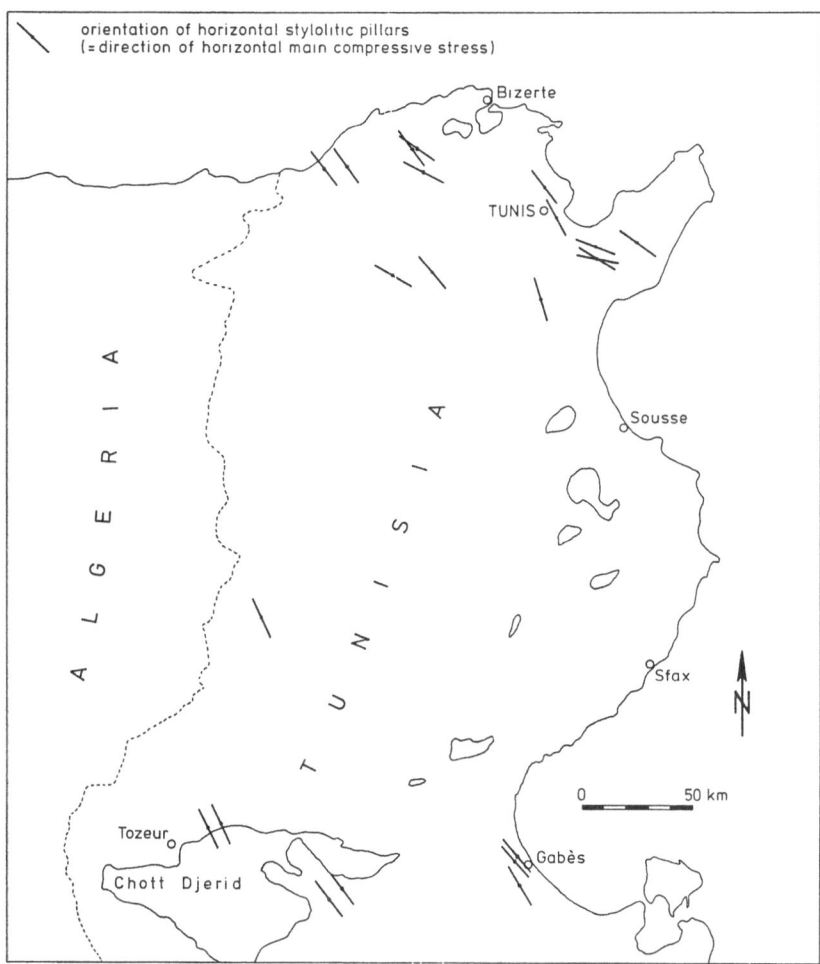

Fig. 2. The younger set of horizontal stylolites is NW-SE-oriented and was formed from the Upper Miocene until Recent. The orientation of the H-stylolites is consistent with the direction of σ_{Hmax} of the stress field that developed during the contemporary NW-SE-approach of Africa and Europe

The directions of σ_{Hmax} and σ_{Hmin} deduced from the study of H-stylolites and calcite-filled veins and the locations are shown in Figs. 3 and 4. The age relation of the NE−SW- and NW−SE-oriented H-stylolites could not be determined with certainty. The intersecting calcite veins revealed an older NE−SW- and a younger NW−SE-trend.

The youngest rocks with H-stylolites of the NE−SW-direction are Eocene nummulitic limestones in northern Cyrenaica. Thus, they were formed

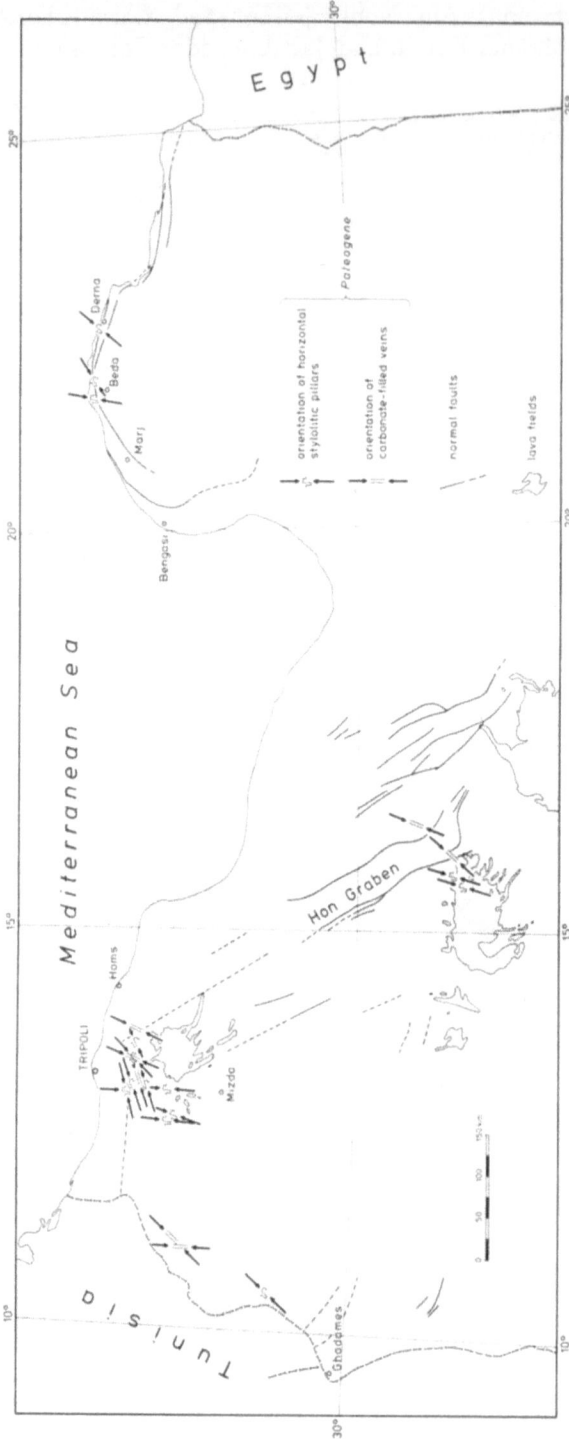

Fig. 3. The NE-SW-trend of the H-stylolites and calcite-filled veins is consistent with the maximum horizontal stress (σ_{Hmax}) of a stress field that was active from the Upper Cretaceous to the end of the Oligocene and had a stress climax from the Middle to Upper Eocene. The contemporary evolution of the Sirte graben and horst structures is in contrast to the Palaeogene stress field

after the deposition and lithification of the Eocene sediments. The younger NW—SE-oriented H-stylolites have been found in the micritic matrix of Quaternary breccias east of Al Giosc in the Jefara plain. No NE—SW-directed H-stylolites have been observed in limestones younger than Eocene.

3. Palaeogene and Neogene Stress Regimes

In Fig. 3, the older set of H-stylolites and calcite-filled veins is shown. It has a NE—SW-trend and was formed during the Middle to Upper Eocene stress climax of a large stress field that influenced major parts of the western Mediterranean and the neighbouring African and European platform areas from the Upper Cretaceous to the end of the Oligocene. The better timing of these stress events has been obtained by structural studies in Spain (Schäfer, 1978). The Libyan results are considered analogously to those of Spain and Tunisia. The locations and the orientation of stress-induced small-scale structures in the Libyan platform carbonates as shown in Fig. 3 are in contrast to the dilatation of the Sirte basin and the formation of the Sirte graben and horst structures at the same time. The maximum subsidence in the southern Hon graben occurred contemporaneously (Palaeocene and Eocene, Klitzsch, 1970) with the formation of H-stylolites and calcite-filled veins indicating a NE—SW-compressive stress that should prevent the graben blocks from subsiding.

Fig. 4 shows the younger Neogene to Recent generation of H-stylolites and veins. It has a NW—SE-trend and was formed after the Burdigalian, most likely during a tectonic stress climax in the Pliocene. Many NW—SE-directed gaping fissures have been used by upwelling lava during the Pliocene, particularly in northern Tripolitania. Ade-Hall et al. (1975 b) dated the Garian volcanics to be Upper Miocene and Pliocene in age (2—6 m. y.). Piccoli (1970) found an older sequence of basalts in the same area which were 53.5—29 m. y. old. Again the locations and the orientation of H-stylolites and calcite-filled veins, as shown in Fig. 4, are in contrast to the evolution of the Sirte graben and horst structures. The Neogene to Recent stress field with its NW—SE-oriented maximum compressive stress and normal to this direction its minimum compressive or maximum tensile stress should have enhanced the rift process in the Sirte area. But at the beginning of the Neogene the subsidence of the Sirte grabens decreased or was even terminated (Klitzsch, 1970).

4. Plate Tectonic Implications

From the bulk tectonic inventory of the western Mediterranean it is concluded that the European-African approach during the late Mesozoic and during the Cenozoic was accomplished in two major phases. One was active from the Upper Cretaceous to the end of the Oligocene with a climax during the Middle and Upper Eocene, and the second major event occurred from the Upper Miocene to the Recent with a climax during the Upper Pliocene.

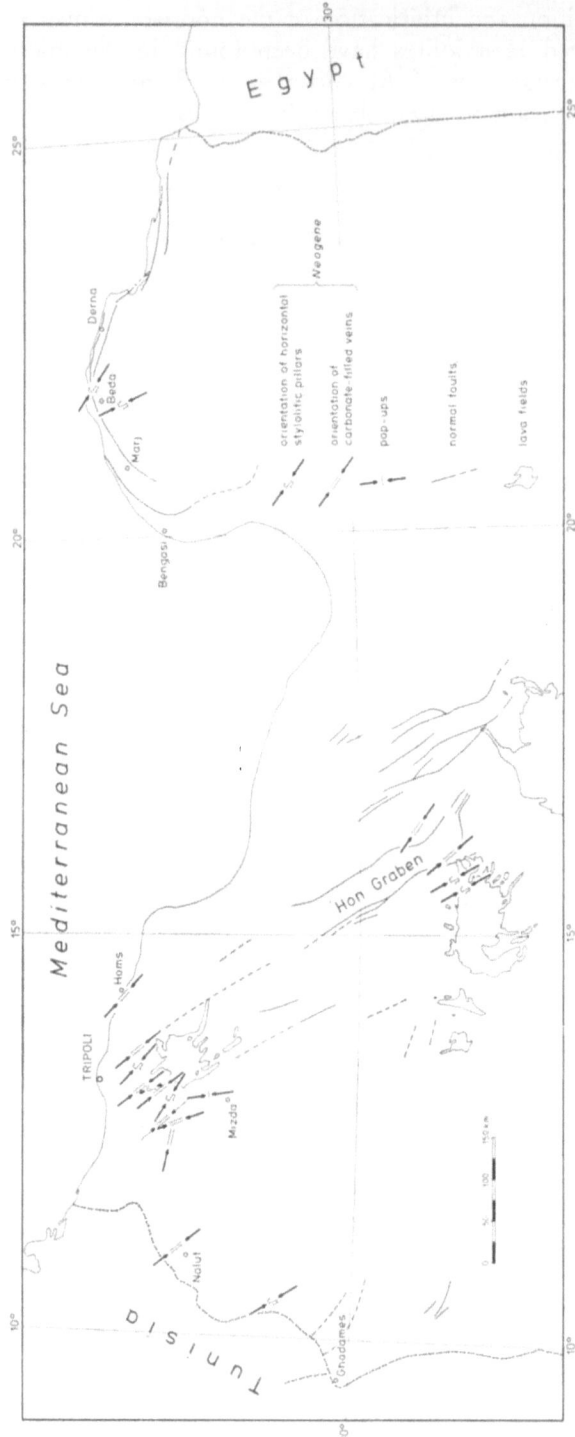

Fig. 4. The Neogene to Recent generation of H-stylolites and calcite-filled veins has a NW-SE-orientation and was formed after the Burdigalian. Again σ_{Hmax} and σ_{Hmin} are in contrast to the decreasing Neogene subsidence of the Sirte grabens

Some of the prominent structural units of northern and central Africa formed during the Upper Cretaceous-Palaeogene-NE – SW-approach of Africa and Europe are shown in Fig. 5 (left side). It is suggested that in the Upper Cretaceous the NE-drifting African plate underwent a continent-continent collision with its northwestern continental extension. The continental crust behaved differently under collision. There was the rigid Adriatic indenter which hit the European continent during that time (pre-Gosau phase) and was responsible for further major folding and thrusting during the Palaeogene ("meso-alpine" orogeny, Trümpy, 1973). This rigid indenter caused a NE – SW-directed horizontal compressive stress to the NW of the continental African plate together with a less rigid or deformable part which was folded and shortened during a pre-Neogene tectonic phase (de Sitter, 1952; Richert, 1971).

The same stress regime may have extended further south causing the closure of the Benue trough and the folding of its sedimentary fill (Olade, 1975).

The NE-section of the African continent east of the Adriatic promontory and its southern extension from Tripolitania to the Chad basin were contemporaneously subjected to dilatation (Sirte grabens, Fig. 5, left side). Thus, the NW-African plate was separated during the Cretaceous from the larger SE-African plate and limited in the south by the Lower Cretaceous opening of the Benue trough and in the east by the Upper Cretaceous rifting in the Sirte basin. At a time shortly after, the Benue trough was closed again (Olade, 1975). The Chad basin subsided during the Mesozoic and connected the Benue trough with the Sirte graben system.

The NW-African subplate during the Upper Cretaceous and the Lower Palaeogene was surrounded by a convergent plate boundary in the north, two divergent plate boundaries in the NE and SW, and by a transform fault system that extended from the Mid-Atlantic ridge to the Chad basin.

Burke and Dewey (1974) discussed the two-plate-theory of Africa and believed that during the Lower Cretaceous a sharp boundary existed along the Benue trough and a diffuse intracontinental boundary toward the NE. In Fig. 3, left side, it is shown that the divergent plate margin in the NE is distinctly limited by the westernmost graben (Hon graben) of the Sirte graben system and its southeastern prolongation, and that only the Chad basin, where transform and divergent movements intersect, shows a diffuse boundary.

Which plate tectonic processes have created those two subplates during the Upper Cretaceous and the Palaeogene? I suggest that after the collision with Europe the NE-drifting African plate reduced its drift velocity from the Upper Cretaceous to the Palaeogene. But only the NW-part was held back while the large SE-African subplate continued drifting to the NE. Both African subplates were subjected to different plate deformations. The NW-African subplate deformed under compression and became smaller (compressive deformable plate), and the SE-African subplate deformed under extension (extensive deformable plate) and broke into more subplates (Arabia, E-Africa).

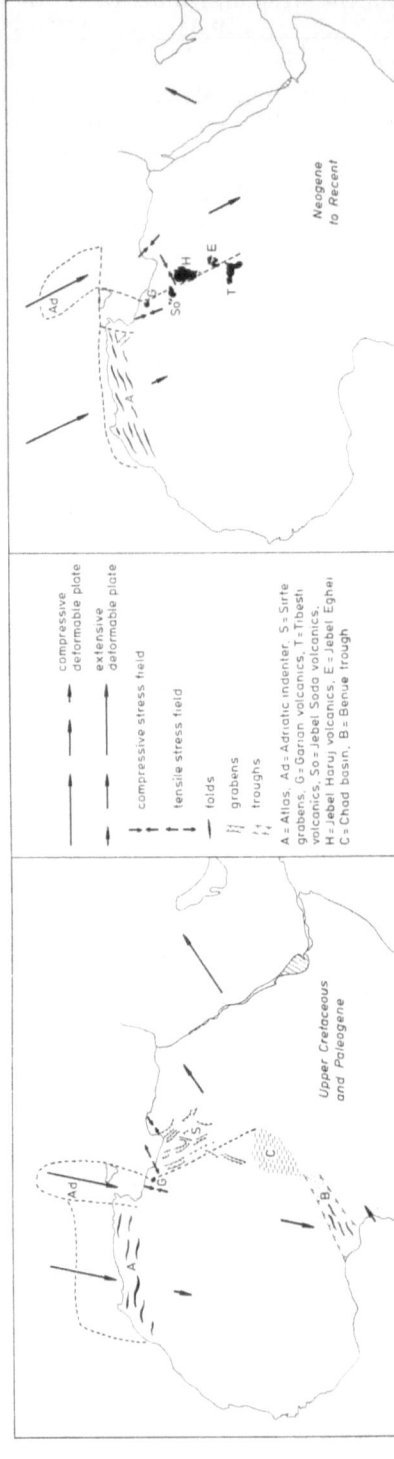

Fig. 5. Left side: During the Upper Cretaceous-Palaeogene-NE-SW-approach of Africa and Europe the Adriatic-African promontory was indenting Europe and caused a NE-SW-directed compressive stress to the NW-African subplate which was deformed by compression. The larger SE-African subplate continued drifting and was deformed by extension

Fig. 5. Right side: An ancient subplate boundary was volcanically reactivated under the NW-SE-directed σ_{Hmax} of the Neogene stress field

From the Neogene to the present time northern Africa seems to be compressed in the NW−SE-direction due to the contemporary African-European approach in that direction. Some of the prominent features which developed during that period are shown in Fig. 5, right side. The ancient subplate boundary was reactivated mainly during the climax of compressive stress in the Pliocene. Some of the volcanic rocks which result from this reactivation have been dated by Ade-Hall et al. (1974, 1975 a, 1975 b). They range from Miocene to Recent in age and erupted predominantly during the Pliocene. Also the Red Sea rift was reactivated during the Pliocene (Girdler and Styles, 1974) as a further result of enhanced compression in the NW−SE-direction and a dilatation normal to it.

Acknowledgements

This work was partly supported by the National Oil Corporation, Tripoli, S. P. L. A. J., and partly by a grant (GF 32510 X) of the National Science Foundation, comprising a special foreign currency grant from the Office of International Programs and a dollar grant from the Earth Science Section of the Division of Environmental Sciences.

References

Ade-Hall, J. M., Reynolds, P. H., Dagley, P., Musett, A. E., Hubbard, T. P., Klitzsch, E.: Geophysical Studies of North African Cenozoic Volcanic Areas I: Haruj Assuad, Libya. Can. Earth Sci. *11*, 998—1006 (1974).

Ade-Hall, J. M., Reynolds, P. H., Dagley, P., Musett, A. E., Hubbard, T. P.: Geophysical Studies of North African Cenozoic Volcanic Areas II: Jebel Soda, Libya. Can. Earth Sci. *12*, 1257—1263 (1975a).

Ade-Hall, J. M., Gerstein, S., Gerstein, R. E., Reynolds, P. H., Dagley, P., Musett, A. E., Hubbard, T. P.: Geophysical Studies of North African Cenozoic Volcanic Areas III: Garian, Libya. Can. Earth Sci. *12*, 1264—1271 (1975b).

Bernoulli, D., Jenkins, H. C.: Alpine, Mediterranean and Central Mesozoic Facies in Relation to the Early Evolution of the Tethys. In: R. H. Dott and R. H. Shaver (eds.), Modern and Ancient Geosynclinical Sedimentation, Soc. Econ. Palaeontol. Mineral. Spec. Publ. *19*, 129 (1974).

Burke, K., Wilson, J. T.: Is the African Plate Stationary? Nature *239*, 387—390 (1972).

Burke, K., Dewey, J. F.: Two Plates in Africa During the Cretaceous? EOS, Trans. Am. Geophys. Union, *55*, 4, 445 (1974).

Channell, J. E. T., Horváth, F.: The African/Adriatic Promontory as a Paleogeographical Premise for Alpine Orogeny and Plate Movements in the Car-patho-Balkan Region. Tectonophysics *35*, 71—101 (1976).

Cocozza, T., Schäfer, K.: Cenozoic Graben Tectonics in Sardinia. Rend. Sem. Fac. Sci. Univ. Cagliari, 145—162 (1974).

Coney, P.: Cordilleran Tectonic Transition and Motion of the North American Plate. Nature *233*, 462—465 (1971).

Girdler, R. W., Styles, P.: Two Stage Red Sea Floor Spreading. Nature *247*, 7—11 (1974).

Klitzsch, E.: Die Strukturgeschichte der Zentralsahara. Geol. Rdsch. *59*, 459—527 (1970).

Olade, M. A.: Evolution of Nigeria's Benue Trough (Aulacogen): A Tectonic Model. Geol. Mag. *112*, 575—583 (1975).

Petters, S. W.: Mid-Cretaceous Paleoenvironments and Biostratigraphy of the Benue Trough, Nigeria. Geol. Soc. Am. Bull. *89*, 151—154 (1978).

Piccoli, G.: Outlines of Volcanism in Northern Tripolitania (Libya). Bull. Geol. Soc. Ital. *89*, 449—461 (1970).

Pitman III, W. C., Talwani, M.: Sea-Floor Spreading in the North Atlantic. Geol. Soc. Am. Bull. *83*, 619—696 (1972).

Richert, J.-P.: Mise en evidence de quatre phases tectoniques successives en Tunisie. Note Serv. Geol. Tunisie *34*, 115—125 (1971).

Schäfer, K.: Tectonic Evidence for Horizontal Compressive Stresses Within the European and North American Plates. EOS, Trans. Am. Geophys. Union *55*, 4, 443 (1974).

Schäfer, K.: Consistent Crustal Stresses Around the Western Mediterranean During the African-European Approach. Rapp. Comm. int. Mer Médit. *24*, 7a, 45—46 (1977).

Schäfer, K.: Geodynamik an Europas Plattengrenzen. Fridericiana *23*, 30—46 (1978).

Schäfer, K.: *In Situ* Strain Measurements in Libya. This volume (1980).

Sclater, J. G., Hellinger, St., Tapscott, Ch.: The Paleobathymetry of the Atlantic Ocean from the Jurrassic to the Present. J. Geol. *85*, 509—552 (1977).

Sitter, L. U. de: Plissement croisé dans le Haut Atlas. Geologie en Mijnbouw *14*, 277—282 (1952).

Trümpy, R.: The Timing of Orogenic Events in the Central Alps. In: K. A. de Jong and R. Scholten (eds.), Gravity and Tectonics, 229—251 (Wiley), New York (1973).

Address of author: Prof. Dr. Karlheinz Schäfer, Institut für Geowissenschaften der Universität, D-8580 Bayreuth, Federal Republic of Germany.

Rock Mechanics, Suppl. 9, 159—172 (1980)

Rock Mechanics
Felsmechanik
Mécanique des Roches
© by Springer-Verlag 1980

Microstructures and Stresses
in Naturally Deformed Peridotites

By

Y. Gueguen and **M. Darot**

With 6 Figures

Abstract

Investigations of the microstructures in naturally deformed peridotites from massifs, basalts and kimberlites have been conducted to identify the deformation mechanisms and to determine the stresses which have produced the plastic flow. Four microstructural parameters can be related to stress: dislocation radius of curvature, free dislocation density, dislocation wall spacing and recrystallized grain size. Theoretical and experimental data on these paleo-stress indicators are discussed. It is concluded that dislocation density is the most reliable one. New experimental data are presented for this parameter. In the present state of our knowledge, only lower bounds on stresses can be estimated. Values of 200 bars are obtained for basalt nodules, 400 bars for kimberlite nodules, 700 bars for the Lanzo massif.

Introduction

Determination of paleostresses in rocks from microstructure is a technique recently introduced by Goetze (1975). Many microstructural data are presently available. In particular, olivines from peridotite massifs and xenoliths have been investigated in details by various authors (Mercier and Nicolas, 1975; Boullier and Nicolas, 1975; Boudier, 1978; Green, 1976; Guéguen, 1977). Using such data, Mercier et al. (1977), Coisy and Nicolas (1978) have inferred stress profiles in the lithosphere. However it appears that these stress estimates rely on theoretical considerations and experimental calibrations which both appear to be unsatisfactory. Nicolas (1978) pointed out that one should not use such techniques without caution and Guéguen (1979 a) has shown that microstructures do not reflect the original deformation conditions in a number of cases. Indeed the "geopiezometer" question must be critically discussed using the most recent and the most accurate experiments in order to define to which extent "piezometers" are reliable and which bounds on stresses can be derived from them.

0080-3375/80/Suppl. 9/0159/$ 02.80

I. Microstructures and Stresses

There are four potential geopiezometers: dislocation radius of curvature, dislocation density, dislocation wall spacing, recrystallized grain size. The first one reflects a very local stress: it can be useful for determining stress in microdeformation experiments, for instance in-situ deformation in the electron microscope, but it is not suitable for determining a mean stress in a rock. Moreover it applies only to perfect dislocations in isotropic crystals, which probably is not the case in olivine (Poirier and Vergobbi, 1978; Guéguen, 1979 b). Consequently we will restrict our discussion to the other three parameters.

1. Free Dislocation Density

The free dislocation density, ϱ, during steady-state creep is proportional to the square root of the applied stress (Weertman, 1968):

$$\sigma = \alpha \mu b \varrho^{0.5} \tag{1}$$

where b is the active Burgers vector, μ the shear modulus and α a constant of order unity. This has been experimentally verified in a number of materials (Barret and Nix, 1965; Takeuchi and Argon, 1976). It can easily be rationalized from elementary dislocation theory. An isolated dislocation exerts a long range stress $\sigma \sim \mu b/r$ (see Friedel, 1965). Dislocations in a crystal are at a mean distance R so that any dislocation experiences a stress of about $\mu b/R$. Since the dislocation density is $\varrho = R^{-2}$, it follows that the mean internal stress is $\sigma_i \sim \mu b \varrho^{1/2}$. If internal and applied stresses are equal, $\sigma = \alpha \mu b \varrho^{1/2}$. Although the exponent 0.5 is predicted by Eq. (1), it is very common in pratice to observe a large number (Nicolas and Poirier, 1976).

Three important problems limit the application of this potential paleostress indicator: (1) measurements of ϱ are not always accurate, (2) the original microstructure has been modified in a number of cases by annealing or late deformation, (3) flow regimes must have been the same (high temperature steady-state creep) for calibration experiments and natural deformation. This last restriction is strengthened by the fact that natural deformation is concerned with polycrystals but accurate experimental calibrations are obtained with single crystals. Ashby (1970) has shown that "geometrically necessary" dislocations are produced during deformation in the first case. Their density is strain- and grain-size dependent. However if deformation takes place at high temperature, these dislocations re-organize into walls and are no longer free.

Experimental calibrations of (1) for olivine have been published by Kohlstedt and Goetze (1974), Kohlstedt et al. (1976), Durham et al. (1977), from creep experiments performed at high temperatures on single crystals (Table 1). Zeuch and Green (1979) have published comparable results, although in that case experiments were performed on polycrystals in the Griggs apparatus. Stresses were measured with an accuracy of a few bars in the first case. The error bars are much larger in the second one. Techniques used for counting dislocations have not been discussed by these authors, although it is a major problem. Hirsch et al. (1965) have drawn attention to the fact that results on dislocation densities may differ markedly when counting methods are not identical.

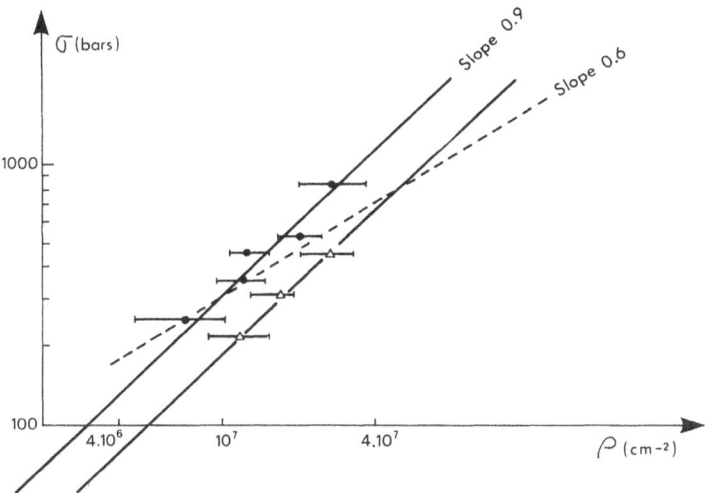

Fig. 1. Stress (σ) — dislocation density (ϱ) diagrams:
dotted line: Durham et al. (1977); solid lines: this work

Our data were obtained from creep experiments performed in a high temperature dead-load machine on olivine and forsterite single crystals, between 1400^0 and 1600^0 C. The atmosphere was controlled by an H_2-CO_2 gas mixture. Stresses are known with an accuracy of a few bars. Dislocation densities were counted systematically with the same method. Dislocations were stained using the techniques described by Kohlstedt et al. (1976) and Jaoul et al. (1979). The use of staining techniques increases largely the reliability of the data since the totality of the dislocation microstructure can be observed and many dislocations are counted. A representative area of 0.5 mm × 0.5 mm was chosen for measurements in the central part of each sample. 10 pictures, each of them covering an area of $80\,\mu \times 60\,\mu$ were taken for counting at the nodes of a grid covering the chosen area. Dislocations close to the perpendicular to the thin section plane were counted by their intersections with this plane: $\varrho = 2\,N/S$ if N is the number of intersections and S the area. Dislocations close to the thin section plane were counted by their intersections with a random network of lines. In that case $\varrho = 2\,N/L \times h$ if N is the number of intersections, L the length of the network and h the depth of focus ($h = 2\,\mu$) (cf. Hirsch et al., 1965). In many cases, the thin section plane was the main glide plane so that most of the dislocations were close to it. Histograms are shown in Fig. 2. The error bar on ϱ is the standard deviation. The procedure outlined above corresponds to counting about 5000 dislocations for each sample.

Two distinct laws (Fig. 1) are obtained for the two different glide systems (010) [100] and (001) [100], corresponding to two groups of experiments. In the first case the crystal is compressed along the $[110]_c^*$ orientation

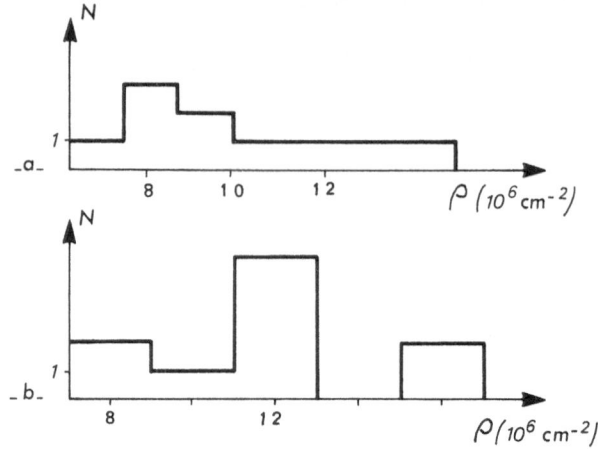

Fig. 2. Histograms of dislocation densities in two samples
a) Olivine single crystal deformed at $\sigma = 368$ bars. b) Peridotite xenolith from kimberlite

so that there is no resolved shear stress on (001) and in the second case the crystal is compressed along the $[101]_c$ orientation so that there is no resolved shear stress on (010). For the same resolved shear stress the dis-

Table 1. Experimental Calibrations for the Different Geopiezometers
(σ in Kb, ϱ in cm^{-2}, d and dg in micrometers)

	This work	Durham et al. (1977)	Mercier et al. (1977)	Post (1973)	Raleigh and Kirby (1970)
ϱ	$0.9 \times 10^{-7} \times \varrho^{0.9}$ $\leq \sigma \leq$ $1.6 \times 10^{-7} \times \varrho^{0.9}$	$\sigma = 2 \cdot 10^{-5} \varrho^{0.6}$			
d		$\sigma = 10\, d^{-1}$	$\sigma = 115\, d^{-1}$		$\sigma = 190\, d^{-1}$
dg			$\sigma = 40\, d^{-0.8}$	$\sigma = 19\, d^{-0.7}$	

location density is higher in the first case. In both cases the exponent is 0.9 whereas Durham et al. (1977) suggested an exponent of 0.6 (Table 1). With the restrictions mentioned above, it is possible to estimate paleostresses as discussed in section II.

---

* The subscript "c" refers to an imaginary arbitrary lattice.

2. Dislocation Wall Spacing, d

In all materials deformed at high temperatures, a large number of dislocations re-organize themselves in low energy arrays. These "bound" dislocations form low to high angle subboundaries (walls) within the grains. Weertman (1968) suggested that the mean spacing d between subboundaries is related to the stress:

$$\sigma = K \mu b d^{-1} \tag{2}$$

where K is a numerical factor of about 100. However there is no clear theoretical justification for Eq. (2). It is likely to be related to the production of geometrically necessary dislocations (Ashby, 1970), which, in turn, depends on strain and grain size. Such dislocations accommodate local strain gradients. If deformation takes place at high temperature, the geometrically necessary dislocations re-organize into low energy walls. The σ-d relationship is probably dependent on parameters such as grain size and strain.

Experimental calibration for olivine have been published by Durham et al. (1977) and Mercier et al. (1977). Raleigh and Kirby (1970) have reported data which cannot be compared with others since these authors considered only subboundaries visible between crossed polarizers (Table 1). Although data of Mercier and Durham incorporate the totality of subboundaries, there is a large discrepancy between them. This may reflect the fact that several σ-d relationships exist. It may also be the result of large experimental errors: data of Mercier were obtained with the Griggs apparatus, in which case the stress is not accurately measured. Experiments of Durham et al. are more accurate but limited to a narrow stress range (100—500 bars). Moreover our results for similar experiments suggest strongly that many of the subboundaries observed by Durham pre-existed the deformation, as discussed below.

Our data support the view that dislocation walls result from geometrically necessary dislocations. No walls are created during homogeneous plastic flow of single crystals whereas (100) tilt walls, and in some cases (001) tilt walls are a feature typical of naturally or experimentally deformed polycrystalline olivine (Green and Radcliffe, 1972; Guéguen, 1979 a). Olivine single crystals used by Durham or by us have the same origin (San Carlos, U. S. A.). They have an original microstructure which consists of many dislocation walls. We believe that most of the walls observed after deformation of such samples were remnants of the former walls. Similar experiments have been conducted with single crystals of synthetic forsterite. Their original microstructure is that of a moderate dislocation density with no walls. Strain as high as 40% has been achieved at low stresses. The corresponding microstructure shows evidence of walls only on the very edge of the sample. In this area, deformation is much less homogeneous (Fig. 3). The mean spacing between walls is locally 12 μ, a value different from that predicted by Mercier (500 μ) or Durham (50 μ) for the same stress (215 bars).

Considering the above-mentioned theoretical and experimental problems related to Eq. (2), it does not seem possible to use it for determining paleostresses, at least in the present state of our knowledge.

3. Recrystallized Grain Size, dg

Recrystallization, the generation of new grains, is a process which is driven by the stored dislocation energy. Sellars (1978) has discussed the microstructural changes during dynamic recrystallization. He suggested that there is an empirical relationship between stress and recrystallized grain size:

$$\sigma = \frac{A}{d_g} n \tag{3}$$

where n is a constant value ($0.5 \leq n \leq 1$). Sellars pointed out that A and n probably depend on the impurity-content. Subboundaries are likely to be the main sites for nucleation of recrystallization in two distinct ways. They can transform into boundaries by progressive misorientation ("in situ" recrystallization) or they can be nuclei for new grains. Guillopé and Poirier (in press) have shown that both these processes are active in halite. In this case several dg-σ equations are obtained, depending on stress, temperature and purity conditions. These two distinct mechanisms are also known to take place during natural deformation of olivine (Poirier and Nicolas, 1975). Consequently, the use of Eq. (3) for paleo-stresses determinations appears to be potentially misleading, unless the mechanisms of recrystallization and the impurity content are identical in nature and laboratory.

Experimental calibrations of (3) have been reported for olivine by several authors (Post, 1973; Mercier et al., 1977) from high stress experiments in the Griggs apparatus (Table 1). Such experiments are unsatisfactory for two reasons: (a) the stress is not known with an acceptable accuracy, (b) the range of stresses investigated is high (1—10 kb) compared to natural conditions.

Durham et al. (1977) have been unable to recrystallize single crystals of olivine, even at large strains (40%). Our results are similar except for one experiment. This difference in behavior between polycrystals and single crystals is comparable to that reported above for dislocation walls. The driving force for recrystallization is the increase in stored energy which results of the accumulation of geometrically necessary dislocations. Plastic deformation of single crystals is homogeneous so that there are no geometrically necessary dislocations, hence no walls and no recrystallization. At very large strains, however, deformation begins to be less homogeneous on the edge of the samples. In this area some walls appear (see 2 above) and recrystallization takes place by progressive misorientation of the walls. However it remains very limited: at $\varepsilon = 27\%$ only one neoblast is present.

Fig. 3. Inhomogeneous deformation (a and b) and homogeneous deformation (c)
a) Forsterite single crystal strained 40%; subboundaries (arrows) are visible in polarized light on the edge of the sample. $\sigma = 215$ bars. b) Same sample at higher magnification: [100] screw dislocations and (100) tilt walls ["(100) organization"]. Walls are indicated by arrows. Plane of the thin section = (010). $\sigma = 215$ bars. c) Same (010) plane in another sample homogeneously deformed at $\varepsilon = 10\%$. $\sigma = 326$ bars. In contrast to b), dislocations are mainly edges (b = [100]) and there is no wall

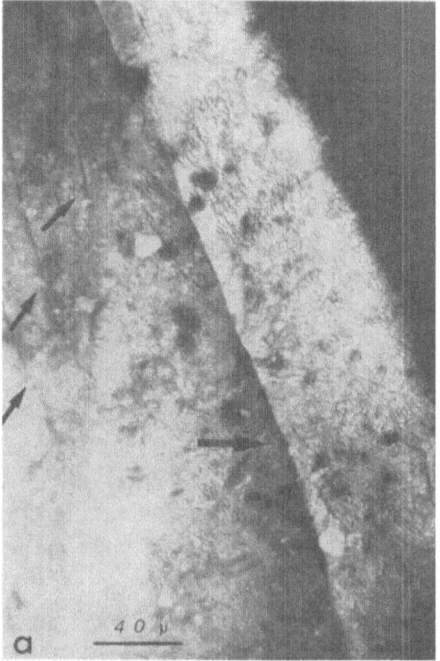

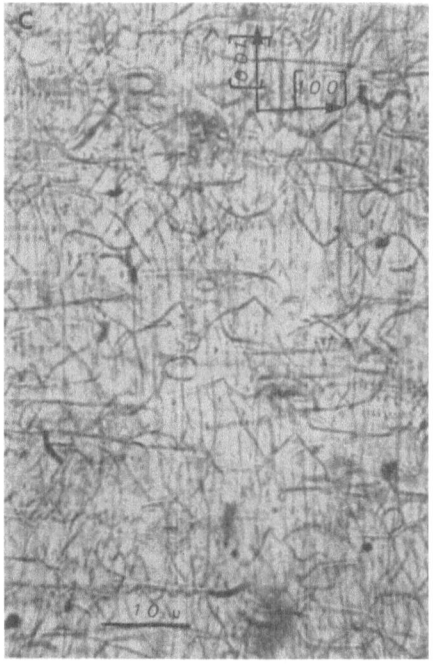

Fig. 3

Its grain size (80 μ) is very different from that predicted by Post or Mercier's calibrations (540 μ and 700 μ) for the same stress (215 bars).

Again, considering the above-mentioned theoretical and experimental problems, it does not seem possible to use Eq. (3) for paleostress determinations.

II. Application to Peridotites

For the reason discussed in section I, it is not clear that unique σ-d or σ-dg relationship exist. In any case no satisfactory calibrations are available. Consequently the free dislocation density ϱ is the only reliable paleostress indicator. Its application, however, is limited by three problems as already mentioned in I-1.

1. Limits of the Method

The first limit is technical: measurements of ϱ must be done systematically with the same technique. Otherwise estimates could differ easily by a factor 2. The technique described in I-1 has been applied to experimentally

Fig. 4. Peridotite from Lanzo massif (Italian Alps): {110} slip bands (low T) superimposed on "(100) organization" (high T)

deformed samples and naturally deformed ones as well. In this last case, a representative grain is chosen and an homogeneous area of 0.5 mm × 0.5 mm is delimited within it for counting.

The second problem is more serious: is the organization of dislocations present in the sample representative of the original one? A population of free dislocations can easily be modified either by a *small late deformation* ($\varepsilon \leq 0.05$) or by post-tectonic annealing. In the first case the perturbation can be evidenced by the dislocation microstructure itself. For instance the glide systems associated to a late low temperature deformation differ from the high temperature ones. Such a situation occurs frequently in massifs. Fig. 4 shows low temperature {110} slip bands superimposed on high temperature

Table 2. Dislocation Densities and Tilt Wall Spacings in Peridotite Xenoliths and Massifs

	Sample	ϱ (cm^{-2})	d (μ)	σ (bars)
1	Xenolith from Arabic Plateau	6×10^6	178	120— 200
2	Xenolith from Arabic Plateau	2×10^6	229	40— 70
3	Xenolith from Hawaii	10×10^6	49	200— 350
4	Xenolith from Baja California ...	10×10^6	65	200— 350
5	Xenolith from South Africa	10×10^6	44	200— 350
6	Xenolith from South Africa	6×10^6	74	120— 200
7	Xenolith from South Africa	26×10^6	7	450— 750
8	Xenolith from South Africa	16×10^6	12	300— 500
9	Lanzo massif	38×10^6	9	700—1100
10	Antalya massif	32×10^6	7	550— 900
11	Gabbro (Ivrea)	38×10^6	4	700—1100

[100] screws and (100) tilt walls. Guéguen (1979 a) has classified the various organizations of dislocations observed in natural deformation of olivine in six groups (Fig. 5): only four of them represent a high-temperature steady state deformation. *Post-tectonic annealing* leads to a decrease of ϱ if the temperature is high enough. Ricoult (1979) has detected significant variations when $T \geq 1200^\circ$ C. After 3 months at 1200° C, ϱ is locally decreased by one order of magnitude. Peridotite massifs have in general been deformed at temperatures lower than 1200° C so that modification by annealing are likely to be small although the annealing time is so long that this conclusion is not definitive. The reverse is true for xenoliths: the magma which brought them to the surface are at high temperatures ($T \sim 1200^\circ$ C). In any case, the stress inferred from ϱ indicates a lower bound.

The third problem is raised by the possibility that some free dislocations may be "geometrically necessary". A local concentration of such dislocations reflects a high local stress resulting from local non-homogeneous strain. If the temperature is high enough, these dislocations reorganize into walls. This situation is observed in a few xenoliths from kimberlites: in the same grain at 100μ apart, ϱ and d vary from 4.5×10^7 cm^{-2} and 25μ to 1.5×10^7 cm^{-2} and 2μ (group 6 in Fig. 5).

2. Massifs and Xenoliths

Dislocation microstructures have been investigated in about 150 peridotite samples from massifs (Lanzo, Italy and Almklovdalen, Norway) and
xenoliths (basalts and kimberlites). The data presented in Table 2 have been

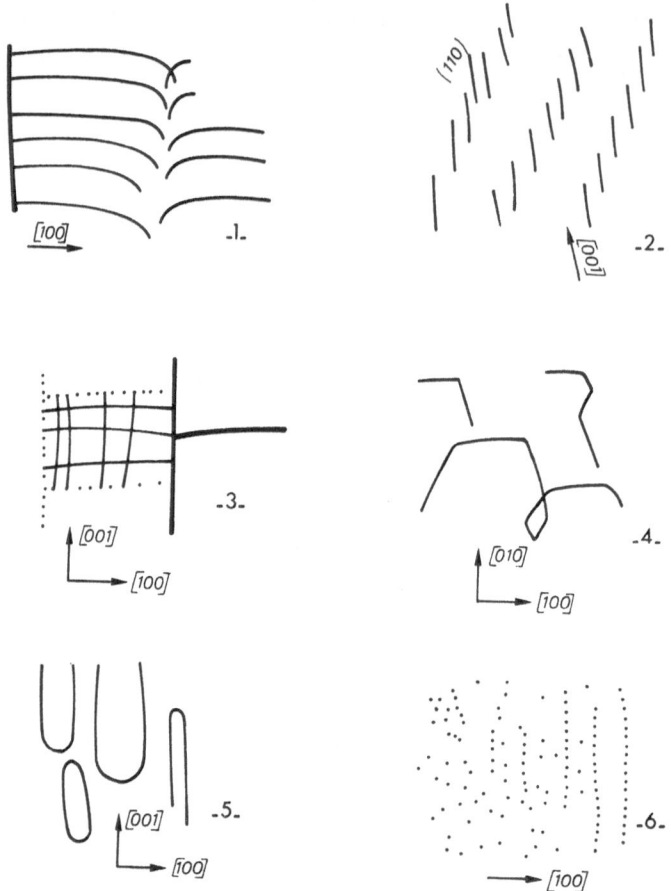

Fig. 5. The six main groups of dislocation microstructures observed in natural deformation

(1) "100" organization (high T) : [100] screws between (100) tilt walls
(2) "001" organization (low T) : [001] screws in {110} slip planes
(3) Polygonization: (100) and (001) tilt walls, (010) twist walls
(4) (001) glide loops (high T): [100] screws and mixed segments
(5) (010) glide loops (high T): [100] screws and long edges
(6) "geometrically necessary" dislocations climbing into walls. Such dislocations accomodate
    the local bending of the crystal

obtained from 11 samples selected on the basis of (1) absence of late deformation, and (2) good homogeneity. Values of ϱ have been determined by
the technique describe above; d values correspond to (100) tilt walls: the

number of counted spacings is ≥ 100. Fig. 6 shows that the (ϱ, d) correlation is very poor, in agreement with what is expected if there is no unique relationship (d, σ). Such data give lower bounds for σ.

The lowest stresses are recorded for xenoliths from basalts (samples 1—4). Minimum stresses range from 40 bars to 350 bars for these xenoliths. Xenoliths from kimberlite (samples 5—8) have been deformed under minimum stresses of 120—750 bars. The highest stresses are recorded for massifs.

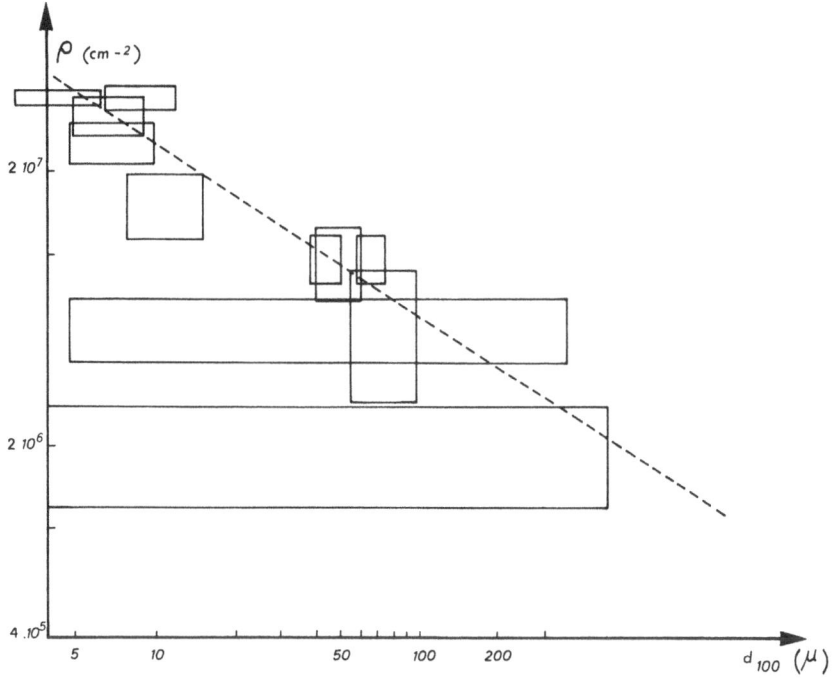

Fig. 6. (ϱ, d) correlation in naturally deformed peridotite. Error bars are standard deviations

In particular the lower bounds for the Lanzo massif and the Ivrea zone are of 700—1100 bars. Geochemical data indicate a different trend for the temperature of deformation: xenoliths from kimberlite have been deformed at 1200^0 C—1400^0 C (MacGregor, 1973; Mercier and Carter, 1975), xenoliths from basalts at 900^0 C—1100^0 C (Coisy and Nicolas, 1978). For the Lanzo massif, the deformation took place from 1200^0 C down to below 1000^0 C (Boudier, 1978). This comparison shows that both σ and T were the lowest for the xenoliths from basalts. Accordingly, they should have experienced the lowest strain rates. If one uses the creep law of Durham et al. (1977) for olivine with $\sigma = 200$ bars, $T = 1100^0$ C and a pressure correction defined by an activation volume $V = 10$ cm^3 mole^{-1} and a pressure of 20 kb, the strain rate $\dot\varepsilon$ is 10^{-11} s^{-1}. Consequently most of our peridotite samples would have been deformed at strain rates faster than 10^{-11} s^{-1}.

Such a value is geologically fast. However the above extrapolations are very uncertain.

In the case of the xenoliths, the deformation probably reflects a local phenomenon. For kimberlite, fast strain rates are the consequence of a rapid intrusion. For xenoliths from basalts, the preferred model would be that of a slow diapiric upwelling (Coisy and Nicolas, 1978). In the case of the Lanzo massif, the high stress recorded may not reflect the main flow regime. It is likely that deformation did not stop completely during the cooling of the massif so that higher stresses may have developed at this last stage, although $\dot{\varepsilon}$ was decreasing to very low values because of the temperature drop. Such an evolution is evidenced in large areas of the massif where the temperature decrease was important enough to induce a change in glide systems (Fig. 4).

Conclusions

Detailed investigations of the microstructures in deformed rocks lead to estimates of paleostresses. However, such estimates are not straightforward and only lower bounds are obtained. Dislocation wall spacing and grain size are not considered to be reliable paleostress indicators. The use of the dislocation density is acceptable on both theoretical and experimental grounds but it is limited by several problems, arising from the fact that a given dislocation population represents only a few percent of strain. Peridotite samples from basalts, kimberlites and massifs have been considered and it is concluded that they have been deformed under minimum stresses of respectively 40—350 bars, 120—750 bars and 700—1100 bars.

Acknowledgements

The authors thank A. Nicolas for discussions and review of the manuscript. This work has been supported by the C. N. R. S. (ERA 547).

References

Ashby, M. F.: The Deformation of Plastically Non Homogeneous Materials, Phil. Mag. 21, *170*, 391—424 (1970).

Barret, C. R., Nix, W. D.: Acta Met. *13*, 1247—1258 (1965).

Boudier, F.: Structure and Petrology of the Lanzo Peridotite Massif. Geol. Soc. Amer. Bull. *89*, 1574—1591 (1978).

Boullier, A. M., Nicolas, A.: Classification of Textures and Fabrics of Peridotite Xenoliths from South African Kimberlites. In "Phys. and Chem. of the Earth" *9*, 467—475. L. H. Ahrens edit., Pergamon (1975).

Coisy, P., Nicolas, A.: Regional Structure and Geodynamics of the Upper Mantle Beneath the Massif Central. Nature *274*, 429—432 (1978).

Durham, W. B., Goetze, C.: Plastic Flow of Oriented Single Crystals of Olivine. I. Mechanical Data. J. Geophys. Res. *82*, 5737—5753 (1977).

Friedel, J.: Dislocations. Pergamon Press. Oxford (1965).

Goetze, C.: Sheared Lherzolites: from the Point of View of Rock Mechanics. Geology 3, 172—173 (1975).

Green, H. W. II: Plasticity of Olivine in Peridotites. In "Electron Microscopy in Mineralogy". 443—464. Wenk edit. Springer, Berlin (1976).

Green, H. W. II:, Radcliffe, S. V. Deformation Processes in the Upper Mantle, Geophys. Mon. Ser. 16, 139—156 (1972).

Gueguen, Y.: Dislocations in Mantle Peridotite Nodules. Tectonophysics 39, 1—3, 231—254 (1977).

Gueguen, Y.: Dislocations in Naturally Deformed Terrestrial Olivine: Classification, Interpretation, Applications. Bull. Mineral. 102, 178—184 (1979a).

Gueguen, Y.: High Temperature Olivine Creep: Evidence for Control by Edge Dislocations. Geophys. Res. Letters 6, 357—360 (1979b).

Hirsch, P. B., Howie, A., Nicholson, R. B., Pashley, D. W., Whelan, M. J.: Electron Microscopy of Thin Crystals. London: Butterworths (1965).

Jaoul, O., Gueguen, Y., Michaut, M., Ricoult, D.: A Technique for Decorating Dislocations in Forsterite. Phys. and Chem. of Minerals 5, 15—19, 1979.

Kohlstedt, D. L., Goetze, C.: Low Stress-high Temperature Creep in Olivine Single Crystals. J. Geophys. Res. 79, 14, 2045—2051 (1974).

MacGregor, I. D.: Petrological and Thermal Structure of the Upper Mantle Beneath South Africa in the Cretaceous. In "Phys. and Chem. of the Earth", Ahrens edit., 9, 455—466. Pergamon (1975).

Mercier, J. C., Carter, N. L.: Pyroxene Geotherms. J. Geophys. Res. 80, 3349—3362 (1975).

Mercier, J. C., Nicolas, A.: Textures and Fabrics of Upper Mantle Peridotites as Illustrated by Xenoliths from Basalts. J. of Petrology 16 (2), 454—487 (1975).

Mercier, J. C., Anderson, D. A., Carter, N. L.: Stress in the Lithosphere: Inferences from Steady State Flow of Rocks. In "Stress in the Earth", Wyss edit., 199—226 (1977).

Nicolas, A.: Stress Estimates from Structural Studies in Some Mantle Peridotites. Phil. Trans. Roy. Soc. London 288, 49—57 (1978).

Nicolas, A., Poirier, J. P.: Crystalline Plasticity and Flow in Metamorphic Rocks. London: Wiley (1976).

Poirier, J. P., Nicolas, A.: Deformation Induced Recrystallization due to Progressive Misorientation of Subgrains, with Special Reference to Mantle Peridotites. J. of Geology 83, 707—720 (1975).

Poirier, J. P., Vergobbi, B.: Splitting of Dislocations in Olivine, Cross Slip Controlled Creep and Mantle Rheology. Phys. Earth Planet. Int. 16, 370—378 (1978).

Post, R. L.: The Flow Laws of Mt. Burnett Dunite. Ph. D. Thesis, U. C. L. A· (1973).

Raleigh, C. B., Kirby, S. H.: Creep in the Upper Mantle. Mineral. Soc. Amer. Spec. pap. 3, 113 (1970).

Ricoult, D.: Experimental annealing of a Natural Dunite. Bull. Mineral. 192, 86—91 (1979).

Sellars, C. M.: Recrystallization of Metals During Hot Deformation. Phil. Trans. Roy. Soc. London A 288, 147—158 (1978).

Takeuchi, S., Argon, A. S.: Steady State Andrade Creep of Single Phase Crystals at High Temperature. J. Mat. Sci. 2, 1542—1566 (1976).

Weertmann, J.: Dislocation Climb Theory of Steady State Creep. Trans. Amer. Soc. Metals, 61—68 (1968).

Zeuch, D., Green, H. W. II: Experimental Deformation of an "Anhydrous" Synthetic Dunite. Bull. mineral. *102,* 180—182 (1979).

Address of authors: Y. Gueguen and M. Darot, Laboratoire de Tectono-physique, Rue de la Houssinière, F-44072 Nantes Cedex, France.

Rock Mechanics, Suppl. 9, 173—192 (1980)

Rock Mechanics
Felsmechanik
Mécanique des Roches
© by Springer-Verlag 1980

Paleo-Stress Fields Around the Mediterranean Since the Mesozoic from Microtectonics. Comparison with Plate Tectonic Data

By

J. Letouzey and **P. Trémolières**

With 9 Figures

Abstract

The method is mainly based on microtectonic measurements in sedimentary rocks: tectonic stylolites, tensional joints, shear planes and faults. Several thousand measurements in different Mesozoic and Tertiary stratigraphic levels from Turkey, Israel, Jordan, Tunisia, Algeria, Morocco, Spain, Portugal, France, Southern England, Southern Italy and Sicily, indicate orientations of local and regional tectonic compressive events, relative chronology and age. These data are compared with the structural evolution of each country.

A sketch of the structural evolution since Upper Cretaceous in the Europe-African-Arabian collision zone is deduced from the compressive events. It gives new elements for the plate tectonic evolution models and continental collisions in the Mediterranean area.

Introduction

The existence and orientation of a present-day compressive stress field around the Mediterranean area is known from in-situ stress measurements and fault-plane solutions of earthquakes. To determine paleo-stress fields, various methods have been used: Models deduced from plate tectonic reconstructions; classical tectonic studies in folded or faulted areas and microtectonic studies in metamorphic rocks. Such structural field studies were mainly concentrated in the Alpine fold belt. Recently, microtectonic analyses have been extended to sedimentary rocks. These studies demonstrated that, in sedimentary basins, rocks recorded also paleo-stress fields thousands of kilometers away from mountain ranges (Arthaud and Choukroune, 1973; Mattauer, 1973).

Geologists and physicists of the French Petroleum Institute worked for several years testing, in the field and with mathematical models, various methods and parameters of structural analysis (Charpal et al., 1974; Quiblier et al., 1978; Quiblier et al., 1979). During the same period, a geological and geophysical synthesis of the Mediterranean was attempted

0080-3375/80/Suppl. 9/0173/$ 04.00

(Biju-Duval et al., 1978, 1979). Structural analysis using microtectonic measurements appeared quickly as a powerful method, bringing a new kind of structural data, especially in the Mediterranean and Alpine continental collision area.

These field studies were carried out in Turkey, Israel, Tunisia, Sicily and France. Studies are now in progress in Jordan, Southern Italy, Algeria, Morocco, Spain and Portugal. They will be completed in the next few years. Thus, the data we present are not final and conclusions are provisional.

Method

By studing the deformations in the field, we try to determine the orientation, the chronology and the age of local and regional shortenings. The local shortening we determine is parallel to the local orientation of the maximum compressive stress. These shortenings are a consequence of compressional tectonic events and the orientations give a good picture of the paleo-stress field. The main measures in sedimentary rock are (Fig. 1): orientation and dip of layers, tectonic stylolite axes, slickensides on wrench, normal and reverse faults, and tensional joints.

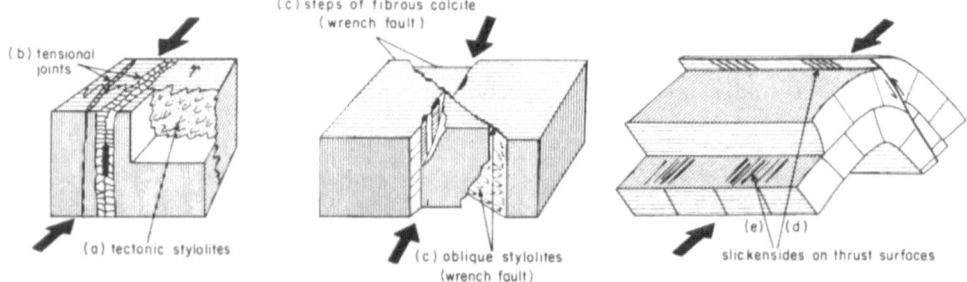

Fig. 1. Microtectonic elements

a) Stylolite axes, when they are perpendicular to the stylolite plane; b) Parallel tensional joint family, only when calcite crystals are perpendicular to joint surfaces; c) Dextral and sinistral strike-slip faults; d) Normal or reverse faults; e) Sliding of competent layers. On shear planes (c, d, e) we make measurements only when oblique stylolites or steps of fibrous calcite give the direction of movement. All measurements made in one site are reported on a Wulff stereogram and interpreted in terms of local shortenings. When all the elements agree in defining one or several directions of shortening, we use strike-slip faults to define precise bounds of the stress sectors (cf. Fig. 2 to 5)

In contrast to other methods, we only measure shear planes when the accurate displacement of the two blocks can be established. It is the only method for knowing whether there is one or several shortening phases which affected the rocks; and if there are several, we are able to give these directions of shortening. By measurements in rocks from different ages we define chronology and age of each phase (Fig. 2). Field observations and mathematical models show that the orientation of the compressional stress depends partly on the nature and the geometry of the geological body and partly on the presence of heterogeneities and discontinuities, particularly large faults. Along these faults, we note frequently a reorientation of stress. Thus, to obtain the best definition of a local stress field, we work on small rock-volumes (a quarry for example), and in a quiet structural zone far

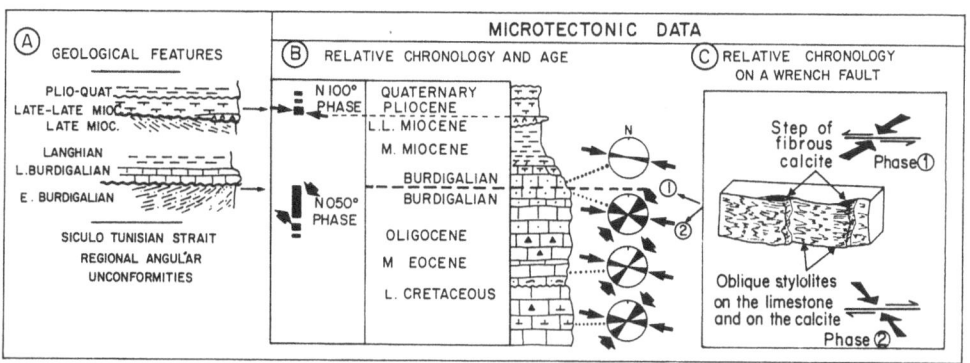

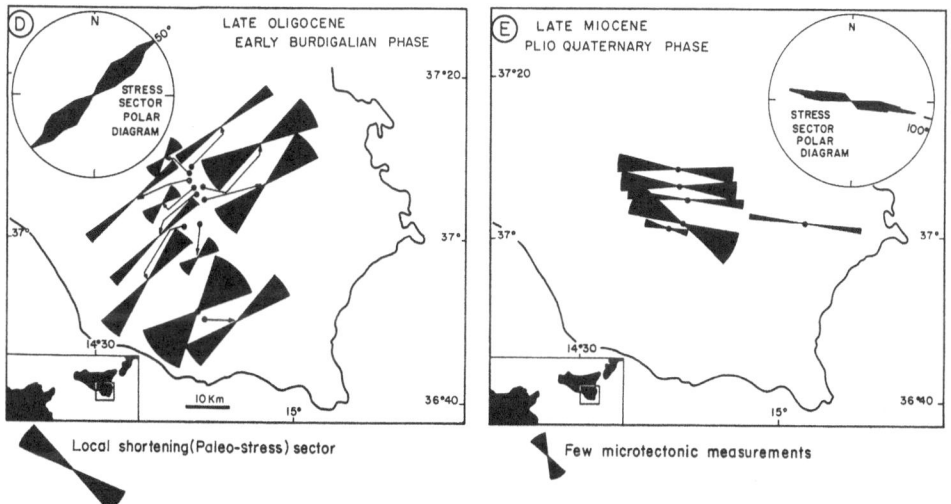

Fig. 2. Dating of regional shortening phases: Example in South Sicily

Four independent criteria are used to define relative chronology between shortening phases and their age: 1 — Relative chronology is sometimes possible when two microtectonic elements are superimposed on the same plane (ex Fig. 2C). 2 — When working on rocks of different ages, some regional orientations of shortening are not observed in the youngest rocks (ex. Fig. 2B). So we have another element of relative chronology and also a close estimate of the age of the termination of a regional shortening phase. 3 — Field experience shows that, except near accidents, paleo-stress orientation is relatively constant or the variation is continuous (ex Fig. 2D—2E). Thus, it is generally possible to correlate local directions of shortening at one site where the chronology is well defined to the nearby sites. 4 — The age of the paroxysm of each phase is determined by the geological framework

(Fig. 2A): foldings, unconformities etc. . . .

In favourable conditions it is thus possible to determine the orientation and the chronology between regional phases; the age of the paroxysm and the release of the stress. But it is very difficult to know the beginning and the duration of each phase from large faults. In each region we try to work in many sites, first in rocks of the same age, then in rocks of various ages.

From the measurements in the same formation we obtain the fossil regional stress field, and the eventual local or regional reorientations; from

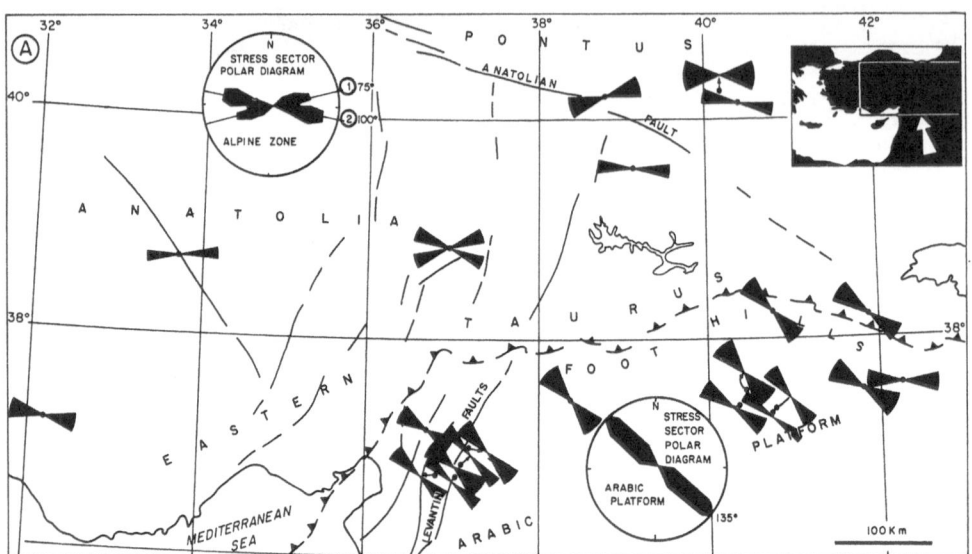

Fig. 3. Shortening phases in Eastern Turkey. Paleo-stress orientation

Fig. 3 A. Late Cretaceous phase

This regional phase affected all the older geological formations up to the Upper Maastrichtian and perhaps a part of Paleocene rocks. It corresponded to the closure of the Tethys mesozoic ocean as proved by Maastrichtian ophiolotic nappes and to the continental collision north of the Arabian Platform (Fig. 9 A). The tectonic paroxysm age is Maastrichtian but shortening continued after thrusting of the ophiolitic nappes. The few data we have on the Alpine zone show first a N 075° direction which turns to N 100°. If these directions are confirmed, they are distinct from the N 135° north Arabian average direction. This could correspond to a change in orientation of the stress due to the indentation process or to sinistral rotation of the Anatolian microplate during collision

Fig. 3 B. Late Eocene — Early Oligocene phase

The second phase was pre-Middle Oligocene in age. Paroxysm probably corresponds to the Late Eocene — Early Oligocene tectonic event well known in the Pontids, Taurus and northern Cyprus mountains

The 20° change in orientation of the stress between the western and the eastern boundary of the north Arabian platform was compared with a three-dimensional mathematical model (Quiblier et al., 1978). This change seems mainly due to the geometry and orientation of the northern boundary of the Arabian plate. The geometry of the blocks also influenced the stress orientations of the following phases (Fig. 3 C, 3 D)

Fig. 3 C. Late Miocene phase

The third phase affected all the geological formations until the most recent one we studied, which was Tortonian in age. The paroxysm age was Late Miocene but shortening could have begun during the Middle Miocene. We observed large changes in orientation of the stress field in different structural areas. In the North Anatolian Fault area two shortening directions affected Miocene rocks. We have no elements of chronology between these two directions. The indentation process north of the Arabian plate caused folding, thrusting, and breaking of continental blocks

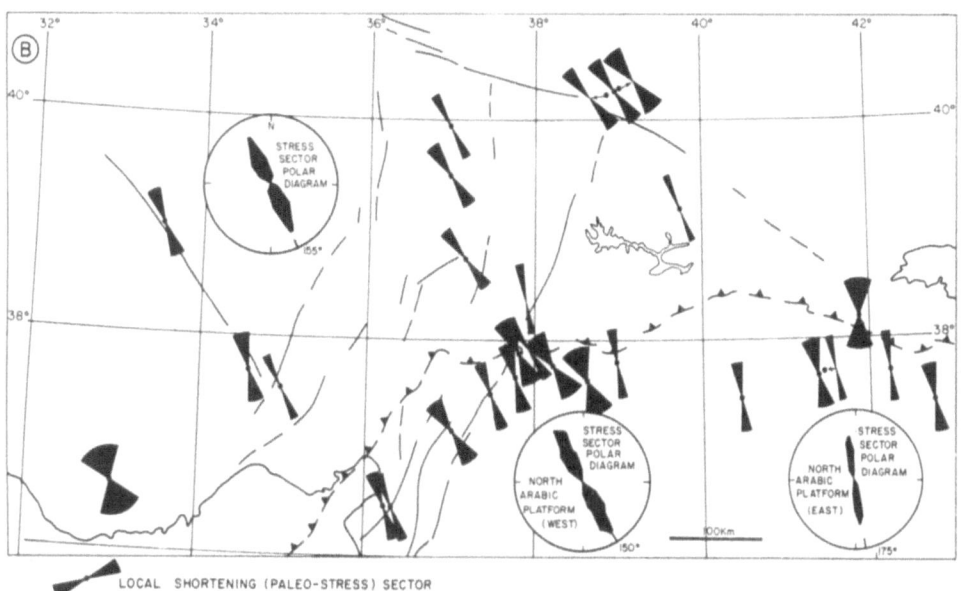

Fig. 3 B

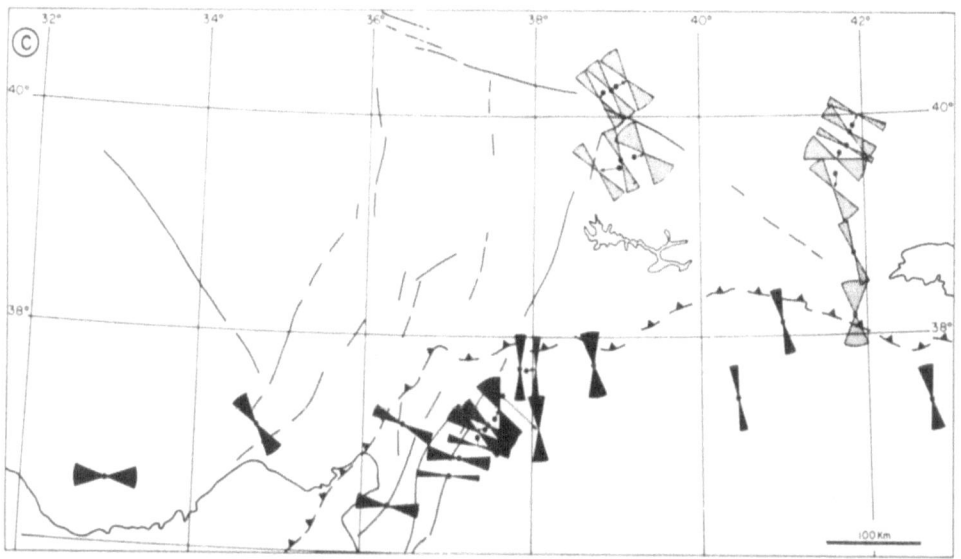

Fig. 3 C

the measurements in different formations, we obtain the evolution of the stress field during geological time.

Although the method is well adapted to the study of shortening (or compression), it is not useful for the study of movements during tensional

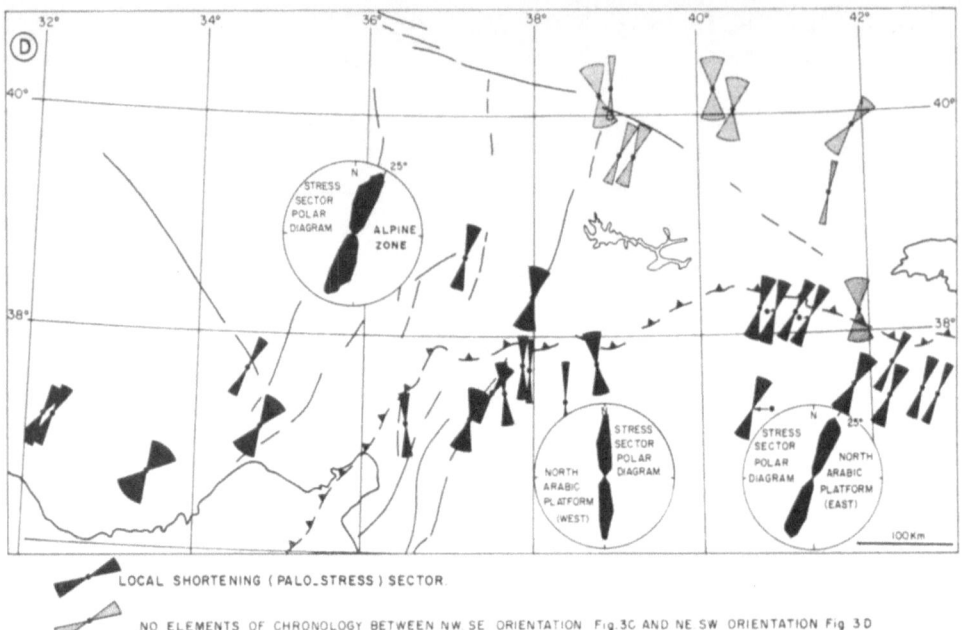

Fig. 3 D. Plio-Quaternary phase

This shortening affected the same formations as the former one, but relative chronology places it after. It is the continuation of the Miocene event after breaking of continental blocks. Faulting, folding and thrusting continued north of the Arabian platform. Large continental wrench faults, for example the Levantine and North Anatolian Faults, led movements of continental blocks (McKenzie, 1978). These movements are still active, with the same direction, as demonstrated by earthquakes

phases. First during the bending of layers by compressional deformation, we can e. g. see local sites with tension at the top of anticlines. Second, in a discontinuous body affected by tension, the only microtectonic elements we can measure are normal faults. All the boundary planes of each block can have a dip-slip component and it is not easy to interpret them in terms of orientation of "regional tensional phase". In this paper we apply the words "tectonic phase" only to regional shortening proved by microtectonic elements and considered as resulting from compressional phases.

Regional Data

In the various countries the quantity of measurements is largely a function of age, nature and quality of the rocks outcropping. Deformation zones are not uniform and are not always at the same place during and between each regional shortening phase. Thus, very often, measurements show only one local shortening phase, even if several regional phases affected the area. Some regional data are exposed on Figs. 3 to 7. Fig. 9 summarizes the main results and conclusions.

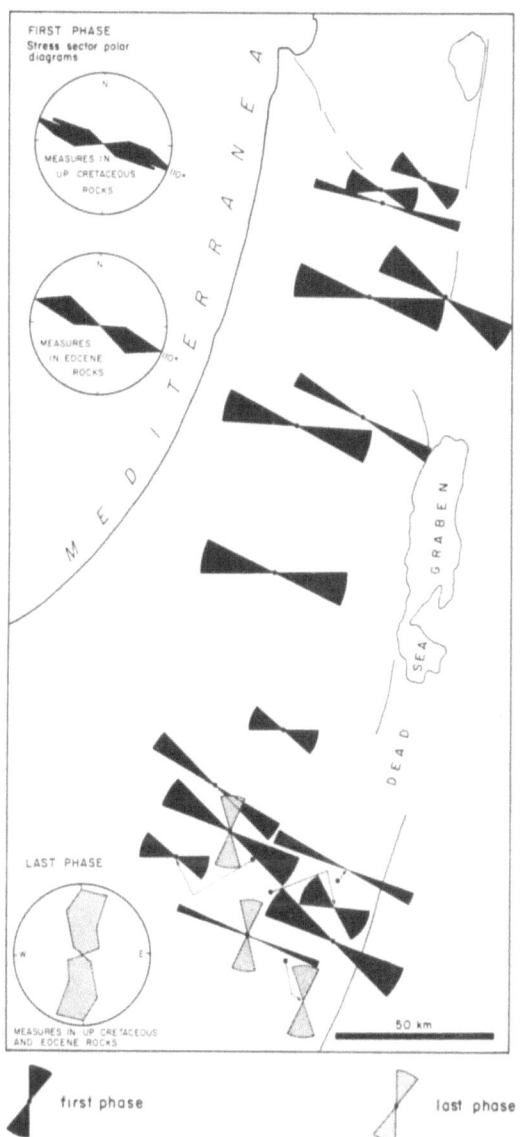

Fig. 4. Shortening phases in Israel. Paleo-stress orientations

As shown by polar diagram the same two paleo-stress directions are found in Cretaceous and Eocene rocks. Microtectonic elements of chronology show that the main phase N 110⁰ is prior to the N-S one. In Israel, folding affected Eocene limestones. Fold axes are consistent with a N 110⁰ shortening. On the other hand, opening of the Dead Sea Rift is inconsistent with this direction. This N 110⁰ phase could have begun in the Late Cretaceous and continued during the Eocene, with precisely the same orientation. But a very similar orientation of the stress is found near the Levantine margin in Turkey during the Late Miocene phase (Fig. 3 C). In our opinion this N 110⁰ phase could be Miocene like the third phase observed in Turkey (Fig. 9 D). The slight N-S phase is consistent with the Plio-Quaternary phase we observed in Turkey (Fig. 3 D). It induced left lateral movement along the Levantine Fault, with the possibility of opening of the Akaba Gulf and Dead Sea Rift (Fig. 9 E)

12*

Eastern Turkey, North of the Arabian Platform (Fig. 3)

The area studied is devided into various structural units. Geological formations range from cretaceous to Late Miocene. We collected more than 2000 microtectonic measurements in about 115 sites.

Israel (Fig. 4)

Measurement sites are mainly in Cretaceous and Early to Middle Eocene limestones. We collected about 450 microtectonic measurements in 17 sites.

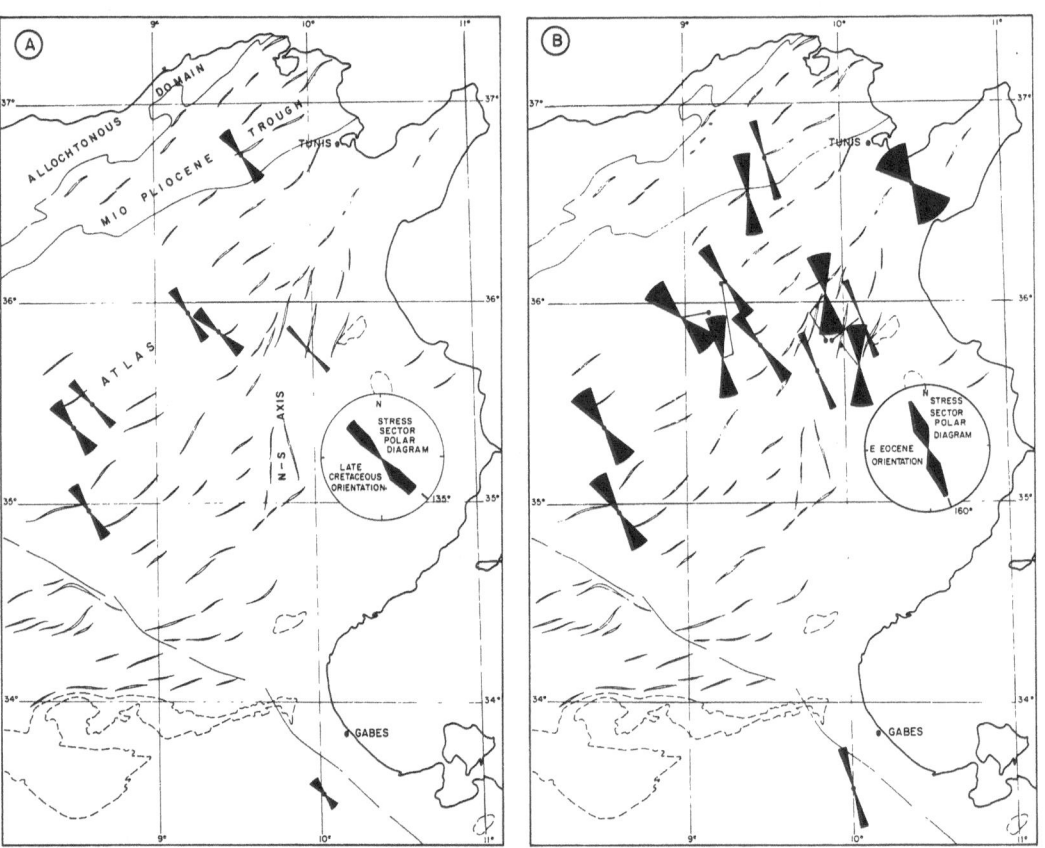

Fig. 5. Shortening phases in Tunisia, Paleo-stress orientation

Fig. 5 A and B. Late Cretaceous to Early Eocene phase

The first phase extended from the Late Cretaceous to the Early or Early-Middle Eocene. The average orientation of the stress is N 140⁰ in Upper Cretaceous rocks (Fig. 5 A) and N 160⁰ in Eocene rocks (Fig. 5 B). We have no microtectonic or geological arguments which could lead us to consider this change in orientation as due to two separate phenomena. The consequences of this shortening are changes in facies, thickenings and folds in the Atlas Mountains

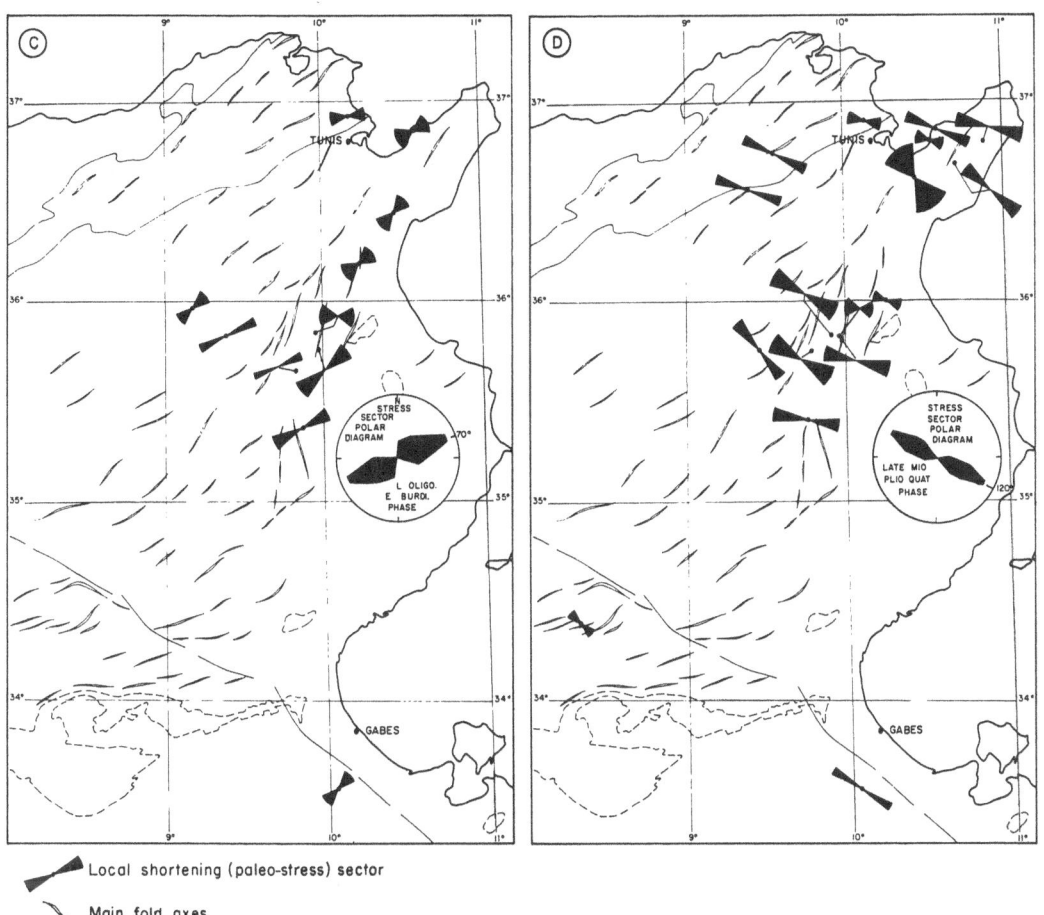

Local shortening (paleo-stress) sector

Main fold axes

Fig. 5 C. Late Oligocene to Early Burdigalian phase

The N 065⁰—N 070⁰ second phase lasted from probably the Late Oligocene to the Early Burdigalian. The late Burdigalian was not affected and lies with an angular unconformity on older eroded formations. The orientation of this phase explains the great structural and paleogeographical changes during this period

Fig. 5 D. Late Miocene — Plio-Quaternary phase

The last N 115⁰—N 120⁰ phase seems to correspond to two paroxysms: the first at the end of the Miocene, the second one during the Quaternary epoch (Burollet et al., 1978). They correspond to the main folding of the Atlas Mountains. From seismicity and recent faulting it seems that present stress orientation is NNW-SSE

Jordan

In Jordan, a NW−SE shortening affected late Cretaceous rocks. The proximity of Israelian measurement sites suggests that this shortening is due to the first phase recorded in Israel (Fig. 4).

Tunisia (Fig. 5)

About 600 microtectonic measures were collected in Middle Cretaceous to Early Miocene rocks. Since Late Cretaceous we have observed three main tectonic phases.

Algeria

A short field study in Mesozoic rocks confirms:

— a N—S shortening direction connected with a late Cretaceous-Early Eocene phase.

— after this phase, a 100° shortening direction (similar to the Tunisian one) and a NE—SW direction but without chronology between them.

Morocco

First data collected in Morocco indicate a N 160—N 170 shortening phase, reoriented in the centre Atlas Mountains perpendicular (N 140) to this range. The paroxysm of this tectonic event seems to be Middle-Late Eocene in age. A Late Cretaceous phase, with paleo-stress orientation different from the Eocene one, was not observed.

In autochthonous geological formations in the north of the country, a N 070 orientation of shortening was observed. Age and chronology of this phase in relation with the Eocene one is not well defined. It could correspond to an Early Miocene event described in the Riff area (thrust of Numidian nappes).

The most important shortening phase is Uppermost Miocene in age, with N 025—N 030 orientation.

Spain and Portugal (Fig. 6)

Important structural work has been carried out in the Northern part of Spain where several thousand microtectonic features have been measured. In the South and the South East part of Spain, ages of the late phases are not clear at the present time.

Southern Italy

The Apulian Plateau and Gargano are considered as autochthonous structural units of the Adriatic plate. We collected 350 measures from Cretaceous to Upper Miocene rocks. Because of the lack of chronology given by the microtectonic elements and of bad distribution of age and location of the sites, chronology and age of the phases are dubious.

A N—S to NW—SE shortening is found in Cretaceous rocks. It could be induced by to a Late Cretaceous to Eocene event. In Northern Apulia we observed in several sites a very clear dextral rotation of the stress, from 132° to N 145°.

A NE – SW shortening, which affects Eocene rocks in the Gargano area, could be attributed to the Early Miocene Apennine phase.

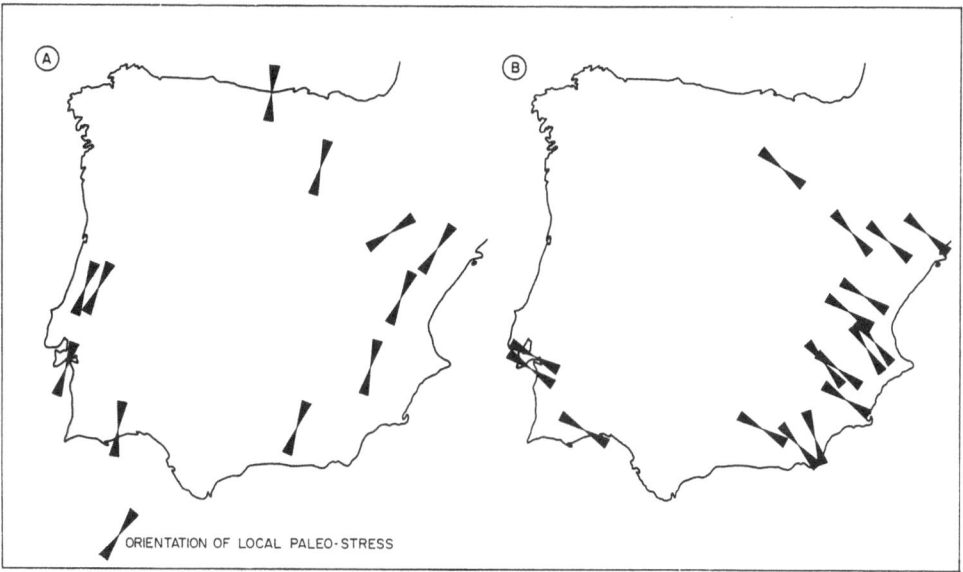

Fig. 6. Shortening phases in Spain and Portugal. Paleo-stress orientation

Fig. 6 A. Late Cretaceous to Eocene phase

A N-S shortening orientation is well represented in the whole area. It could have begun in Senonian time. It is interesting to note a local reorientation of the stress along the large NE-SW tectonic trends (Catalonia Chains, Lisboa Faults)

Fig. 6 B. Post Burdigalian phase

Shortening phases affected some Miocene rocks. But studies in Late Miocene outcrops showed a complicated structuration and additional work is necessary in the South and South-East area. Note the lack of the NW-SE shortening in the North and North West of the peninsula where measurements are abundant

In the same site, Late Miocene rocks are affected by a 140⁰ shortening probably due to a Uppermost Miocene or Plio-Quaternary event.

Sicily (Fig. 2)

The African platform extends to the North of the Ibleo Area, South of Sicily. About 300 microtectonic measurements were collected in Late Cretaceous to Messinian limestones.

If we compare southern Sicily (Fig. 2) and Tunisia (Fig. 5) there is good agreement between the two last phases: we have the same age and the same angle between the orientations of these two phases. But if the angles are similar, the directions of shortening in Sicily correspond to a 15⁰ sinistral rotation with respect to Tunisian directions. This can be explained by a rotation of Sicily during Plio-Quaternary times due to a difference in the rate of extension along the Pantelleria grabens.

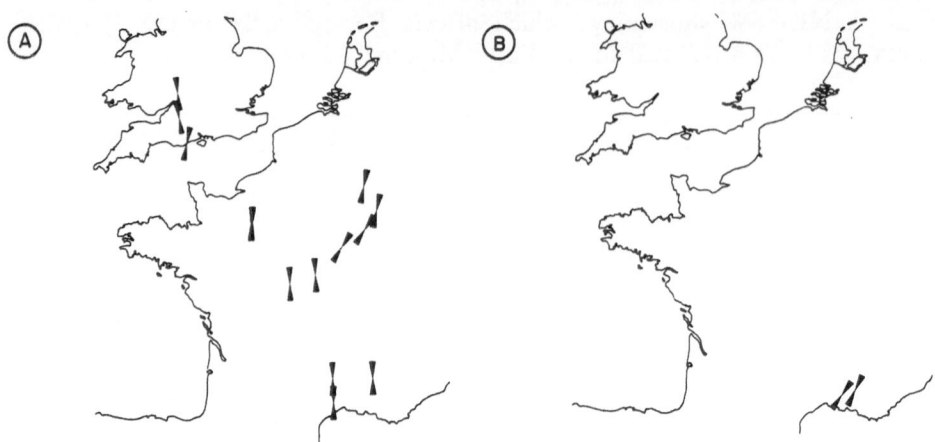

Fig. 7. Shortening phases in France and Southern England. Paleo-stress orientations

Fig. 7 A. Late Cretaceous to Eocene phase

The N-S shortening is well represented in the whole area and sometimes the first Oligocene formations were affected. As in Spain, we note stress reorientation, parallel to the main NE-SW tectonic trends (Burgundy, Cevennes Faults)

Fig. 7 B. Burdigalian phase

A NE-SW compressive event, affecting the Lower Burdigalian limestones, is well observed in the South East of France

Fig. 7 C. Late Miocene to present phase

A third phase occurred after the Late Burdigalian rocks were deposited. It is oriented NW-SE, and only represented in the Eastern and South Eastern parts of France near the Alpine folded zone. In these areas it affected the Tortonian formation. Its age could be Late Miocene to Present

In England, an E-W compression is observed in Jurassic rocks but there is no chronological element with respect to the N-S Eocene one

France and Southern England (Fig. 7)

In France, more than 10 000 measurements have been made. Local studies are numerous and in general the deformation mechanisms are well known.

Conclusions and Discussions

It seems useful to place the compressive tectonic phases deduced from field data in the larger framework of the Europe-African-Arabian collision, and to appreciate the contribution of this method to the understanding of the structural evolution of the Mediterranean and of collision phenomena. A first sketch of the evolution of the Mediterranean area since the Late

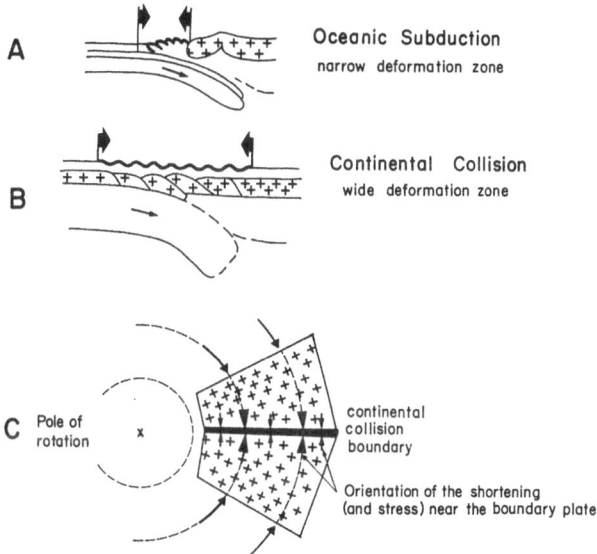

Fig. 8. Shortening phases and Plate Convergence

The shortening phases we measured by structural analysis are a consequence of plate convergence and continental collision phenomena

Fig. 8A. If at least one of the convergent plates is oceanic, the deformation zone is very narrow, as proved by seismic reflection on active margins and zones of superficial seismicity

Fig. 8B. In the case of continental collision, the zone of deformations appears very wide. The consequence of the Europe-Africa collision is presently a zone of seismicity wider than 1000 km. This is a consequence of buoyancy of continental or intermediate crusts (McKenzie, 1978)

Fig. 8C. A very schematic model shows that if the boundary between two convergent continents is perpendicular to the relative movement of the plate, orientation of shortening near the boundary could indicate the direction of the relative movement between the two plates. Thus, the directions we measured by structural analysis could be parallel to the small circles around the pole of rotation of the two plates, and could have fossilized these directions. If we could know the rate of shortening all along the collision boundary during each shortening phase, we would have a very good approximation of the relative movement of the convergent plates

Cretaceous time is presented Fig. 9. It was done to verify whether there was consistency between: paleo-stress measurements, relative movement of the blocks, and zones of collision and deformation. To draw this sketch, the following points were taken into consideration: the geological framework and the necessity of continental or rigid block collisions which caused the compressional events we observed by structural analysis. For each phase we tried to bring the blocks closer together, starting from their previous positions parallel to the measured orientation or regional shortenings (Fig. 8).

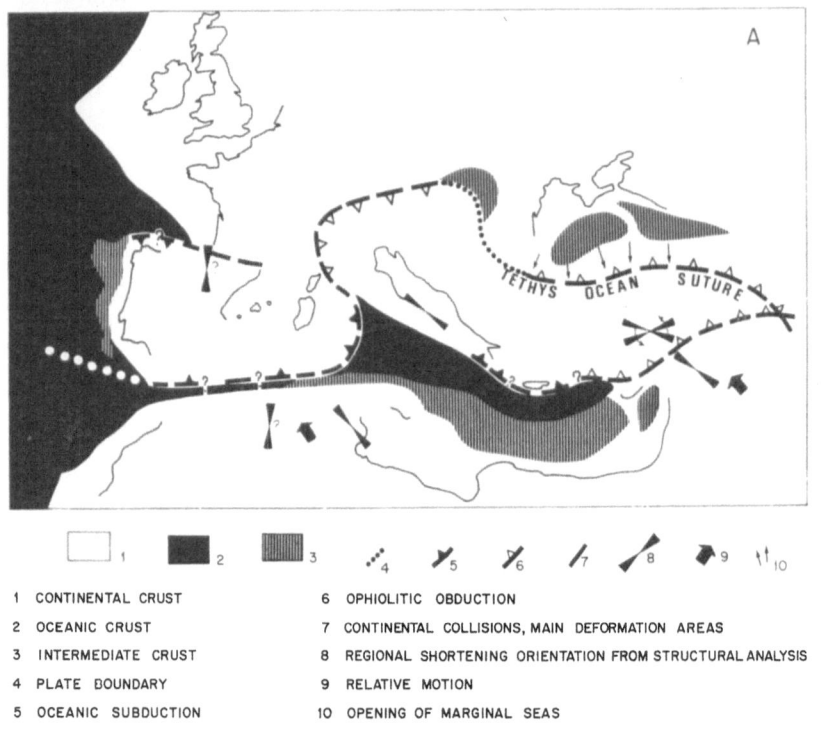

1 CONTINENTAL CRUST			6 OPHIOLITIC OBDUCTION		
2 OCEANIC CRUST			7 CONTINENTAL COLLISIONS, MAIN DEFORMATION AREAS		
3 INTERMEDIATE CRUST			8 REGIONAL SHORTENING ORIENTATION FROM STRUCTURAL ANALYSIS		
4 PLATE BOUNDARY			9 RELATIVE MOTION		
5 OCEANIC SUBDUCTION			10 OPENING OF MARGINAL SEAS		

Fig. 9. Sketch of evolution of the Mediterranean basins and paleo-stress field orientations On all the figures Europe was arbitrarily fixed. Regional paleo-stress orientations were in fossil positions

Fig. 9A. Late Cretaceous

Late Cretaceous shortening corresponds to the closure of the Tethys Ocean as attested by nappes with ophiolitic and deep water deposits in the Eastern Alpine zone, Taurus, Pontus Mountains. Orientations of shortening North of the African (AF) — Arabian (AR) plate were mainly NW-SE with respect to European reference. Some tectonic deformations are described during Upper Cretaceous in the western area, but microtectonic measurements did not show specific orientations different to those observed during Eocene time. Thus, we cannot confirm that shortenings and continental collisions began in the Upper Cretaceous period in that area

The Eastern Mediterranean Sea is the relic of the Southern Mesozoic Tethys (Biju-Duval et al., 1978). This oceanic area could explain the lack of shortening in Sicily and perhaps in Israel during the Cretaceous and Eocene period (absence of continental collision in this zone). NW-SE Syrte graben systems were subsident after this shortening period

Fig. 9 B. Eocene

From the Upper Cretaceous to the Eocene, we observed in several areas changes in orientation of paleo-stress from NW-SE to NNW-SSE. Eocene is one of the main periods of shortening in the whole Mediterranean area. However paroxysms are not synchronous all along the collision zone. The Rhine, the Rhone, the Eastern and Western Sardinia Graben systems subsided after this shortening period

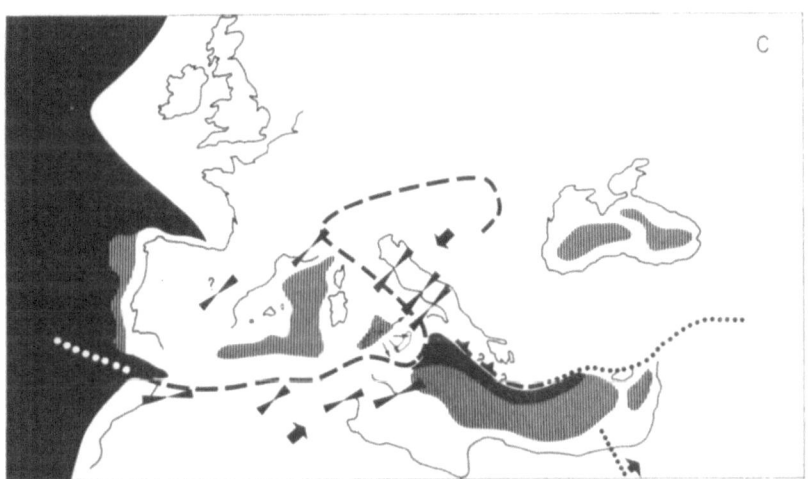

Fig. 9 C. Late Oligocene — Early Miocene

NE-SW paleo-stress orientations observed all around the central Mediterranean area during the Late Oligocene and Early Miocene is completly different from the former ones

In fact, we began with the present position of the continent. We moved back the blocks more or less in parallel to the last orientation of the regional compression we observed. Thus, proceeding with phase after phase, we obtained the relative position of the blocks at each period of time. The

Fig. 9 D. Late Miocene

The second major shortening phase is the Late Miocene one. The rapid change in orientation of the stress in function of the area probably corresponds to: a blocking up of the EU-AF movement; the indentation process north of a continental promontory like Arabia; a breaking of continental blocks and a new orientation of relative movements. This tectonic event which closed the straits between the Mediterranean, the Atlantic, and the Persian Gulf, is probably one of the factors which caused the Messinian salinity crisis

The Panteleria Graben system between Tunisia and Sicily was active after this shortening phase. A difference in the rate of extension in this zone during the Plio-Quaternary can explain sinistral rotation of Sicily with respect to Tunisia, deduced from paleo-stress measurements

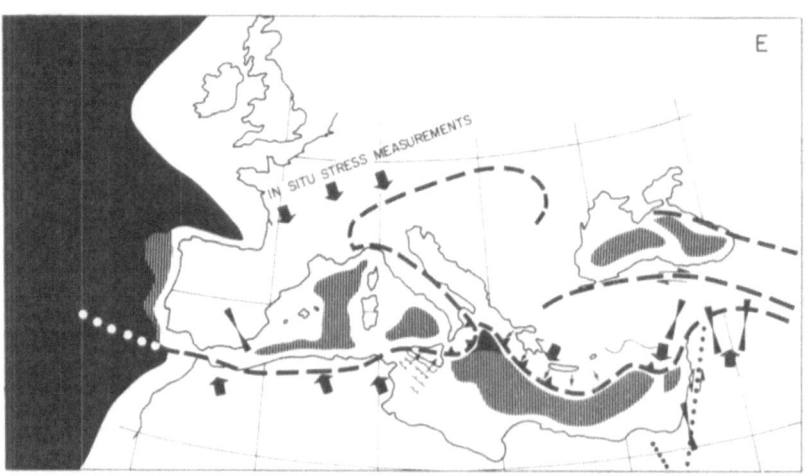

Fig. 9 E. Plio-Quaternary

In Turkey and Israel measurements confirm the movement along the Levantine Fault. In other areas, movements are too recent or too small to be interpreted with microtectonic methods. Fault plane solutions of earthquakes and in-situ stress measurements seem to indicate a more or less N-S European-African relative movement

least definite factor is clearly the amount of shortening and the deformation of the continental blocks.

Although the phases of compression seem synchronous in the whole Mediterranean area, it is not the case for their paroxysm. But to simplify the sketch, the orientations of regional shortening were grouped into five periods (Fig. 9 A, B, C, D, E): Late Cretaceous, Eocene, Late Oligocene-Early Miocene, Late Miocene and Plio-Quaternary.

Measurements taken in Jurassic formations did not show shortening phases previous to the Upper Cretaceous one. It would have been interesting to make this kind of measurement in countries where Jurassic or Early Cretaceous movements are well described: Dinarides and Hellenides (Mercier, 1966; Auboin et al., 1970); Morocco (Mattauer et al., 1977); The Balkans (Dewey et al., 1973).

From the Upper Cretaceous period to the present time, comments about the paleo-stress changes, in relation to the structural evolution of the Mediterranean are shown in Fig. 9.

From the paleo-stress determinations some general remarks arose:

— Newly created small basins with oceanic or intermediate crust, such as the western Mediterranean Basin or the Tyrrhenian Basin, seem to transmit tectonic stresses applied to their boundary and behave like rigid blocks. The same applies to old oceanic basins or margins with thick sediments, such as the Black Sea Basin or the Levantine Basin south of Cyprus. Deep seismic profiles in these basins show few deformed sediments, whereas large deformations appear under their margins and on land all around (Finetti and Morelli, 1973; Biju-Duval et al., 1979; Letouzey et al., 1977).

— Structural analysis shows that the subsidence of graben and the regional shortening phases are not synchronous but there is a close correlation between the orientations of the normal faults along the grabens and the orientation of the regional shortening just before the subsidence, as deduced from structural analysis.

— Deformations at every scale, which correspond to shortening, are observed in intra-cratonic basins, very far away from the collision zone, but they are linked and synchronous with those observed along collision zones. In some cases, microtectonic analysis shows the disappearance of the deformations away from folded zones at various distances.

— During a regional shortening phase, the orientation of paleo-stress was relatively constant on the scale of the studied area, particularly in geologically stable and homogeneous areas. Reorientations of stress were always localized near discontinuities: margins, faulted zones, plate boundaries.

— Despite these local reorientations, age and orientation of paleo-stress measured give a coherent kinematic model of the relative movements of plates and blocks around the Mediterranean Basin since the beginning of the collision.

The data obtained by this structural study bring some new constraints to the previous models obtained by other methods: geological and geophysical studies of the Mediterranean area (Biju-Duval et al., 1978); rigid plastic analogic model (Tapponnier, 1977); paleomagnetism (Sclater et al., 1977) and Atlantic oceanic magnetic anomalies (Pittman and Talwani,

1972; Dewey et al., 1973; Biju-Duval et al., 1977; Auzende et al., 1979, etc. ...).

The age and orientation of the relative movements of the plates and collision events deduced from paleo-stress determinations differ sometimes from the rigid plate kinematic models deduced from Atlantic magnetic anomalies. One can point out particularly the following differences:

— The distance between Northern Arabian, Anatolian and European continents was much smaller during the Late Cretaceous period than suggested in many models because continental collision occurred North and South of Anatolia at that time.

— The orientation of the plate movements during the Eocene time is different and the amount of shortening much more important.

— The NE−SW direction of compression during the Early Miocene was unknown. The width of the area which was deformed, includes a large part of the continents away from the collision zone; the changes of orientations of paleo-stress observed during a certain period of time cannot be explained if plates or microplates are rigid with a constant geometry.

These facts seem to confirm some of the conclusions of the rigid-plastic collision theory (Molnar and Tapponier, 1975). Nevertheless, we did not observe the large variation of maximum compressive stress trajectories inside the continents foreseen in this theory; reorientations of the stresses were mainly localized near the collision zones, and orientations of the paleo-stresses deduced from structural analysis are surprisingly constant on wide areas on continents. Moreover, the very distinct changes of paleo-stress orientations during geological time contradict the progressive change of orientations adopted in the model proposed for the Mediterranean (Tapponnier, 1977).

The schema of structural evolution of the Mediterranean area which we proposed from paleo-stress determinations, is provisional and must be modified after new measurements and if possible a better estimate of the amounts of shortening have been made.

Nevertheless the estimate will probably always be very rough and therefore the relative movement of the plates cannot be determined with precision. We intend to use both, the data from structural analysis which bring new constraints and the movements deduced from Atlantic magnetic anomalies, to obtain a better scheme of the structural evolution of the Mediterranean area.

Acknowledgments

The authors wish to thank all the geologists of IFP: O. de Charpal, B. Biju-Duval, G. Jacquart, L. Montadert, C. Salle, B. Zinszner who participated to the field trips; G. and J. J. Bizon, C. Müller who did the stratigraphic determinations; the specialists of the numerous countries who helped us: P. F. Burollet, A. di Grande, R. du Dresnay, F. Hirsch, S. Kozak, B. Ozer, M. Romeo, G. Suter, Ph. Trouve, and many others we cannot mention here. We also wish to thank Ch. Bois, L. Montadert, M. Poulet for discussing the manuscript and especially R. Kidd for his critical review.

References

Ahorner, L.: Present Day Stress Field and Seismotectonic Block Movements along Major Faults in Central Europe. Tectonophysics 29, 233—249 (1975).

Angelier, J.: Sur l'évolution tectonique depuis le miocène supérieur d'un arc insulaire méditerranéen: l'Arc Egéen. Rev. de Géogr. Phys. et de Géol. Dynamique (2), 29, (3) 271—294 (1977).

Angelier, J., Mechler, P.: Sur une méthode graphique de recherche des contraintes principales également utilisable en tectonique et en séismologie; La méthode des dièdres droits. Bull. Soc. Géol. France (7) 29 (6), 1309—1318 (1977).

Arthaud, F., Choukroune, P.: Méthode d'analyse de la tectonique cassante à l'aide des microstructures dans les zones peu déformées. Exemple de la plateforme Nord-Aquitaine. Rev. de l'Inst. Fr. du Pétrole 27 (5), 715—732 (1973).

Auboin, J., Blanchet, R., Cadet, J. P., Celet, P., Charvet, J., Chorowicz, J., Cousin, M., Rampnoux, J. P.: Essai sur la géologie des Dinarides. Bull. Soc. Géol. Fr. (7) 12, 1060—1093 (1970).

Auzende, J. M., Olivet, J. L., Bonnin, J.: Les données de la cinématique et l'évolution du domaine Méditerranéen occidental (To be published, 1979).

Biju-Duval, B., Dercourt, J., Le Pichon X.: From the Tethys Ocean to the Mediterranean Seas: A Plate Tectonic Model of the Evolution of the Western Alpine System: In "Structural History of the Mediterranean Basins". Edit. Biju-Duval, B. and Montadert, L., Paris, Technip. 143—164 (1977).

Biju-Duval, B., Letouzey, J., Montadert, L.: Structure and Evolution of the Mediterranean basins. In: Hsü, K., Montadert, L., et al., 1978. Initial Reports of the Deep Sea Drilling Project, Vol. 42, Part 1: Washington (U. S. Government Printing Office), 951—984 (1978).

Biju-Duval, B., Letouzey, J., Montadert, L.: Variety of Margins and Deep Basins in the Mediterranean. In: "Geological and Geophysical Invertigations of Continental Margins". Edit. Watkins, J. S., Montadert, L., and Dickerson, P. N., A. A. P. G. Memoir 29, 293—317 (1979).

Burollet, P. F., Mugniot, J. M., Sweeney, P.: The Geology of the Pelagian Block: The Margins and Basins off Southern Tunisia and Tripolitania. In: "The Ocean Basins and Margins". Vol. 4 B the Western Mediterranean. Edit. Nairn, A. E. M., Kanes, W. H., and Stehli, F. G., 331—360 (1978).

Charpal, O. de, Trémolières, P., Jean, F., Massé, P.: Un exemple de tectonique de plate-forme: les Causses majeures (Sud du Massif Central, France). Rev. de l'Inst. Fr. du Pétrole 39 (5), 641—659 (1974).

Dewey, J. F., Pitman III, W. X., Ryan, W. B. F., Bonnin, J.: Plate Tectonics and Evolution of the Alpine System. Geol. Soc. of Am. Bull. 84, 3137—3180 (1973).

Finetti, I., Morelli, C.: Geophysical Exploration of the Mediterranean Sea. Bol. Geofisica Teor. et Appl. 18 (69), 31—65 (1973).

Illies, J. H.: Ancient and Recent Rifting in the Rhine Graben. Geologie en Mijnbouw 56 (4), 329—350 (1977).

Letouzey, J., Biju-Duval, B., Dorkel, A., Gonnard, R., Kristchev, K., Montadert, L., Sungurlu, O.: The Black Sea: A Marginal Basin, Geophysical and Geological Data. In: "Structural History of the Mediterranean Basins". Edit. Biju-Duval, B., and Montadert, L., Paris, Technip, 263—375 (1977).

Mattauer, M.: Les déformations des matériaux de l'écorce terrestre. Paris, Hermann, 493 p. (1973).

Mattauer, M., Tapponnier, P., Proust, F.: Sur les mécanismes de forma-tion des chaines intra-continentales. L'exemple des chaines atlasiques du Maroc. Bull. Soc. Géol. France (7) 29 (3), 521—526 (1977).

McKenzie, D.: Active Tectonics of the Alpine-Himalayan Belt in the Aegean Sea and Surrounding Regions. Geophys. J. R. Astr. Soc. 55 (1), 217—254 (1978).

Mercier, J.: Sur l'existence et l'age des deux phases régionales de méta-morphisme alpin dans les zones internes des Hellénides en Macédoine centrale (Grèce). Bull. Soc. Géol. France (7) 8, 1014—1049 (1966).

Molnar, P., Tapponnier, P.: Cenozoic Tectonics of Asia: Effects of a Con-tinental Collision: Science 189, 419—426 (1975).

Paquin, C., Froidevaux, C., Souriau, M.: Mesures directes des contraintes tectoniques en France septentrionale. Bull. Soc. Géol. France (7) 20 (5), 727—731 (1978).

Pitmann III, W. C., Talwani, M.: Seafloor Spreading in the North Atlantic. Geol. Soc. Amer. Bull. 83, 619—643 (1972).

Quiblier, J., Becquey, M., Poulet, M.: Les modèles mathématiques des déformations peuvent-ils être utiles au géologue? Mémoire du B. R. G. M. 91, 125—135 (1978).

Quiblier, J., Trémolières, P., Zinszner, B.: A Tentative Fault Propagation, Description by Three Dimentional Finite Element Analysis. Tectonophysics (in press, 1979).

Schäfer, K.: Consistent Crustal Stresses Around the Western Mediterranean During the African-European Approach. Rapp. Comm. int. Mer Medit., 24, 7a (Abstr.) (1977).

Sclater, J. G., Hellinger, S., Tapscott, C.: The Paleobathymetry of the Atlantic Ocean from the Jurassic to the Present. J. Geol. 85, 509—552 (1977).

Tapponnier, P.: Evolution tectonique du système alpin en Méditerranée: poinçonnement et écrasement rigide-plastique. Bull. Soc. Géol. France (7) 29 (3), 437—460 (1977).

Address of authors: J. Letouzey, P. Trémolières, Institut Français du Petrole, 1—4 avenue de Bois Préau, F-92506 Rueil-Malmaison, France.

Rock Mechanics, Suppl. 9, 193—199 (1980)

Rock Mechanics
Felsmechanik
Mécanique des Roches
© by Springer-Verlag 1980

Theme 5

Recent Displacements

Vertical Movements in Switzerland

By

E. Gubler

With 1 Figure

Abstract

The first order precision levelling in Switzerland includes about 3000 km levelling lines, which form 18 loops with an average of 220 km each. The first measurements were made from 1903—1925. Since 1943, 60% of the net has been successively
measured a second time. The comparison of these two levellings shows vertical
movements of the earth's crust, since the Alps rise by up to 1.5 mm/year with respect
to the Swiss plateau. An error analysis proves that these movements are in all
probability significant. They are clearly shown in a graph. Also, various local measuring stations have been set up to prove movement specifically along fault lines.

Introduction

In Switzerland work has been in progress for several years to determine vertical movements from the 1st order national levelling. Jeanrichard
first reported on this work in 1972. Since then, there have been further
publications; by Jeanrichard in 1973 and 1975, Pavoni in 1975, Gubler
in 1976 and Kobold in 1977. This contribution should report on the present
work being done in this direction.

The Swiss First Order Levelling

The national levelling net consists of 3000 km of levelling lines. It consists of 18 loops with an average length of 220 km, has 550 km of connecting lines to neighboring countries and includes about 13,000 bench marks,
usually bronze rivets, which are placed in groups of 3 rivets. Wherever
possible, they are embedded in rock, and if this is not possible, they are
anchored in the walls of buildings with good foundations. The average distance between the bench mark groups is about 2—3 km. For the measurement, iron rivets, placed in 200—400 m intervals, are used as auxiliary marks.

0080-3375/80/Suppl. 9/0193/$ 01.40

The net was first measured from 1903—1925 with the best accuracy attainable in those days. From 1914 on, exclusively invar rods were used. Since most of the lines with significant height differences were measured after 1914, they have a good scale. The standard deviation of a one-km back and forth levelling, calculated from the loop-misclosures, is 1.4 mm.

Since 1943, the net is being successively measured a second time, and today, 60% of the net has been remeasured with the same care. Mainly the WILD N3 levelling instrument was used until 1970 and today the WILD NA2 compensator-levelling instrument is being used. The standard deviation, calculated from the loop-misclosure, is around 0.8 mm.

Method to Determine Vertical Movement

It is easy to determine vertical movement if the time interval between two measurements is relatively large compared to the time used for making each measurement. Unfortunately this is not the case for the Swiss levelling net. The first measurements were made over a time period of 23 years, and the second measurements have been going on for 36 years and are far from being finished. For this reason the interval between the two measurements varies between 20 and 74 years, meaning the interval is sometimes shorter than the time period needed for one measurement. Therefore, the movement of the bench marks on which the measurement was interrupted for longer periods of time must not be ignored. There is no information available to show how the movement may have changed in the course of time. In the absence of more exact information, it must be assumed that the movement remained uniform. For this case, Holdahl's model 2 (1975) is best suited. Velocity differences Δv are calculated from the height differences Δh determined for every common section of the first and second levelling using the formula:

$$\Delta v = \frac{\Delta h_2 - \Delta h_1}{\Delta t}$$

whereby Δt is the time interval in years. The velocity difference is expressed in mm/year. Accordingly, the standard deviation $m_{\Delta v}$ can be calculated for each velocity difference using the formula:

$$m_{\Delta v} = \frac{1}{\Delta t} \sqrt{(m_1{}^2 + m_2{}^2) \cdot S}$$

whereby m_1 and m_2 are the estimated standard deviations for 1 km of back and forth levelling in the first, respectively the second levelling, and S is the length of the respective section in kilometers. From these formulas, one can see that the accuracy of the velocity difference depends not only on the accuracy of the two levellings involved, but also increases as the time interval between the two levellings increases. Furthermore, it can be assumed that systematic errors influencing the levellings will disappear in the difference, so far as both levellings are affected in the same way, which is at least partly the case with refraction and other known influences.

It is useful to adjust the velocity differences, calculated according to this method, by using the least squares method. Thereby, uplift and subsidence velocities, refering to an arbitrarily chosen reference bench mark, as well as a variance/co-variance matrix for these velocities, are obtained for all of the investigated bench marks. The standard deviation for all of the velocities as well as for each interesting relative velocity between any two bench marks can be calculated from this matrix. However, it must be re-emphasized, that this adjustment, for lack of more exact information, is based on the hypothesis of uniform movement.

Vertical Movements Computed from 1st Order Levelling

The above-mentioned method was applied to those parts of the national levelling net which have been measured twice. 110 bench marks which seem to be closely bound to the earth's crust were chosen. Most of them are embedded in rock, and the average distance between the bench marks is 10—20 km. A net of 9 closed loops can be formed with the available data so far. For the adjustment, the velocity of a bench mark in Aarburg was assumed to be zero [Pavoni (1975)]. The uplift and subsidence velocities relative to this reference bench mark as well as their standard deviations are calculated and shown in the graphic representation. The lined columns represent the velocities, and the heavy lines on either side of the columns represent the doubled standard deviation, which corresponds to a level of significance of 95%. Uplifts and subsidences are differentiated by the direction of the columns. This graphic representation has the advantage that significant relative movements can easily be distinguished from accidental results. No significant movement relative to Aarburg can be proven if the lined columns are shorter than the heavy lines. If the column is larger, a relative movement is probable. An interpolation of isolines has not been made so far, as the available data is incomplete and leaves too many gaps open. The relative movement between any two bench marks can be easily obtained as the difference between their velocity columns, and their standard deviation can be computed from the variance/co-variance matrix.

Critical Evaluation of Results and Method

Fig. 1 shows that no significant movement can be proven in the Swiss Plateau. In the Alpine region, however, significant uplifts with respect to the Plateau can be ascertained. They reach up to 1.7 mm/year near Brig and near Chur, which is six times the amount of their standard deviation, and are thus in all probability significant. Furthermore, significant differences in the uplift velocities can be found within the Alps. They are probably not due to scale errors, since the largest uplift rates appear at the bottom of the Alpine valleys at 600 m above sea level. With the exception of a few lines measured before 1914 without invar rods, no correlation between the determined uplift velocities and the heights can be found. However, a few reservations must also be added:

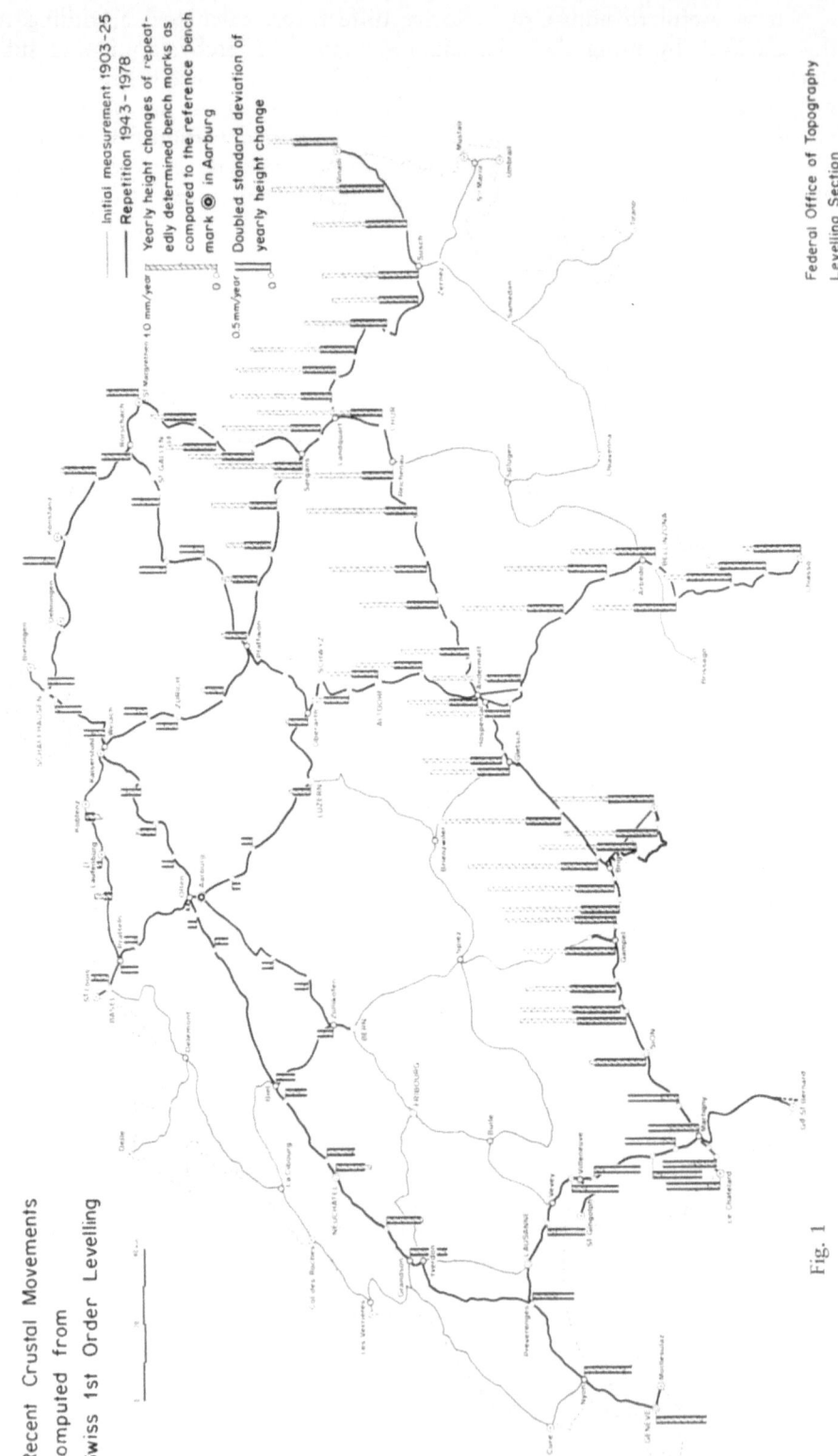

Fig. 1

— Only relative movement can be determined with the available measurements. Therefore, a subsidence of the Plateau would produce the same results. In order to answer this question, other measuring techniques must be taken into consideration.

— The bench marks used in this study were chosen by geodesists. A further study needs to be made which takes into account the geological conditions.

— Every geometric levelling always refers to the equipotential surfaces of the earth's gravity field. Therefore, the ascertained differences include changes of the earth's gravity field, changes which could have been caused by density changes or shifts of masses within the earth. Studies in this direction still need to be undertaken.

— Even though the movements are represented in mm/year, it must be stressed that they are the result of two levellings made at an interval of 20—74 years. The hypothesis of uniform movement may be incorrect.

To this last remark it must be added that there are several indications which show that the hypothesis of uniform movement might be more or less correct. Similar uplifts have been determined by: Levallois (1972) for the French, Senftl (1970 and 1973) for the Austrian and Starzmann (1976) for the Bavarian Alps. Kobold (1977) found a surprising agreement of the above mentioned results to those from a levelling made in the last century, even though the accuracy of the latter was poor. Clark and Jäger (1969) have found comparable uplift rates for the period of the last 30 million years from geo-chronological studies. Pavoni (1975) emphasizes the close association between upliftrates, Bouguer anomalies and crustal stresses in the Alpine area, as derived from the analysis of focal mechanisms. Jaeckli (1958) has ascertained denudation rates for the Alpine region which agree with the determined uplift rates.

On the other hand, the movement of the bench marks during the measurement could partly explain the loop-misclosures in the levellings. They are influenced by the movement of the bench marks during the measurement if it stretched over a longer period of time. However, an unequivocal answer to the question of whether the movements are uniform or not, cannot be obtained from this investigation. Several loops were measured in such a short time that the movement has no consequences. A part of the rest of the loops shows the expected decrease of loop-misclosures. In a Swiss as well as in an Austrian/Swiss loop, the loop-misclosures increase from +21 to +32 mm, respectively from −28 to −50 mm, which could have been caused by an irregular movement of the nodal bench mark in Sargans. But other explanations cannot be excluded.

Local Investigations

Based on geological investigations, special levelling lines were set up which cross well-known fault lines showing signs of possible recent movement, with the aim of proving anticipated movement. Two lines cross the

eastern main fault line in the Upper Rhinegraben near Basel. One loop has
been measured near Andermatt right on the Rhine-Rhone fault line. The
first measurements were made in 1974 and 1973. First results will not be
available before the repeat measurements, planned for 1980—1981, are completed. In 1976, additional geodetic profiles were established in the Gütsch-
Stöckli-Lutersee area on the Rhine-Rhone line.

Final Remarks

The vertical movement determined using the Swiss first order levelling
should not be regarded as a conclusive result. There are still many questions
to be answered, and in this report, a few of the unanswered questions have
been mentioned. A close cooperation between all of the European countries
as well as between the various earth sciences is necessary to bring us closer
to the answers to these fascinating questions.

References

Clark, S. P., Jr., Jäger, E.: Denudation Rate in the Alps from Geochronology
and Heat Flow Data. Am. Journ. Science 267, 1143—1160 (1969).

Fischer, W.: Rezente Erdkrustenbewegungen in der Schweiz. Vermessung,
Photogrammetrie, Kulturtechnik 5-74, 113—115 (1974).

Gubler, E.: Beitrag des Landesnivellements zur Bestimmung vertikaler Krustenbewegungen in der Gotthard-Region. Schweiz. mineral. petrogr. Mitt. 56,
675—678 (1976).

Holdahl, S. R.: Models and Strategies for Computing Vertical Crustal Movements in the United States. Preprint for International Symposium on Recent Crustal
Movements, Grenoble (1975).

Jaeckli, H.: Der rezente Abtrag der Alpen im Spiegel der Vorlandsedimen-
tation. Eclogae geol. Helv. 51/2, 354—365 (1958).

Jeanrichard, F.: Contribution à l'étude du mouvement vertical des Alpes.
Boll. di Geodesia e Scienze Affini 1 (1972).

Jeanrichard, F.: Nivellement et surrection actuelle des Alpes. Rev. Mensura-
tion, Photogrammétrie, Génie rural 1 (1973).

Jeanrichard, F.: Rapport sur l'état actuel des recherches sur les mouvements
verticaux des Alpes, présenté à la XVIe Assemblée générale de l'UGGI, Grenoble
(1975).

Kobold, F.: Die Hebung der Alpen aus dem Vergleich des „Nivellement de
Précision" der Schweiz. Geod. Kommission mit dem Landesnivellement der Eidg.
Landestopographie. Vermessung, Photogrammetrie, Kulturtechnik 4-77, 129—137
(1977).

Levallois, J. J.: Sur la mise en évidence d'un mouvement de surrection des
massifs cristallins alpins. Bull. Géodésique, n. Ser. No. 105, 229—312, Paris 1972.

Pavoni, N.: Report of the Working Group on Recent Crustal Movements.
First Report of Switzerland to International Geodynamics Project, July 1975, 3—17,
Schweiz. Natf. Ges., Bern 1975.

Senftl, E.: Ein Beitrag zum Nachweis rezenter Bewegungen in den Hohen Tauern. Oesterr. Zeitschr. für Vermessungswesen *58*, 41—47 (1970).

Senftl, E., Exner, Ch.: Rezente Hebung der Hohen Tauern und geologische Interpretation. Verh. Geol. B.-A. Wien 2, 209—234 (1973).

Starzmann, G.: Präzisionsnivellement und rezente Vertikalbewegungen der Alpen. Zeitschr. für Vermessungswesen *8* (1976).

Address of author: E. Gubler, Federal Office of Topography, Seftigenstrasse 264, CH-3084 Wabern, Switzerland.

Rock Mechanics, Suppl. 9, 201—212 (1980)

Rock Mechanics
Felsmechanik
Mécanique des Roches
© by Springer-Verlag 1980

Repeated Levelling and Vertical Crustal Movements. Problems and Results

By

N. Höggerl

With 5 Figures

Abstract

The first chapter of this paper deals with the geodetic elements on which our investigations of vertical crustal movements are based. This chapter is followed by a short survey of those levellings carried out by the Bundesamt für Eich- und Vermessungswesen and its predecessor, the Militärgeographisches Institut.

As an identification of vertical crustal movements necessarily requires a high standard of accuracy of measurements, the measurements available will be subject to close scrutiny. In this connection, the various possibilities of determining accuracy will be examined. It will be shown that the standard deviation for 1 km double levelling may vary between 0.5 mm/km (calculated from lines) and 1.9 mm/km (calculated from the loop-misclosure). Then a first presentation of vertical crustal movements in Central and Eastern Austria will follow. It will be based upon the comparison of the measurements carried out in the 1950s and those of the 1970s. This first draft, however, shows a certain uncertainty, which can be seen from the high standard deviation that appears in some regions.

From the extensive measurements available, which even date back to the past century, a detailed study of part of the Wiener Becken will be presented. With reference to the bench marks in Schottwien we find maximum velocities of -1.9 mm/year near Guntramsdorf, which obviously depend on the conditions of the subsoil.

Introduction

It has been tried for some decades now to make use of precise levellings for the determination of vertical crustal movements. Before these levelling results can be used to draw any definite conclusions, they have to be subject to close scrutiny. These investigations are that important because the individual measurements were carried out at different periods. That is to say, that the equipment and the methods have certainly been improved between two periods.

When large levelling nets were set up more than 100 years ago, the standard deviation was 4—5 mm/km. Meanwhile, the levelling accuracy has been constantly increased. Especially during the last few years, a rapid

0080-3375/80/Suppl. 9/0201/$ 02.40

development in this particular branch of geodesy has taken place. New instruments such as the laser interferometer have been introduced (Schlemmer, 1975). In addition to this, various methods for computing vertical crustal movements have been developed (Holdahl, 1975; Ghitau, 1970). These different improvements, however, must not obscure the fact that the levelling accuracy that can be achieved at the moment only reaches approximately 0.4—0.6 mm/km.

Existing Levellings in Austria

The levelling data available in Austria are derived from 4 different periods of measurements. The oldest levellings of first order on Austrian territory were carried out by the Militärgeographisches Institut (MGI), the predecessor of the present Bundesamt für Eich- und Vermessungswesen (BAfEuVW). These measurements date back to the 19th century. They were performed between 1873 and 1895. This part of the net was about 3000 km long and comprised 7 loops.

Between 1920 and 1938 in many Central European countries the old net had to give way to one which was established according to the latest investigations. In Austria, however, only about 800 km of precise levellings were measured at that time. There were several lines but no actual net. The lack of loops had a serious impact upon the control of measurements.

The year 1948 marked the beginning of a new levelling net. It consisted of 57 loops, which were about 7500 km long. These measurements were finished in 1962. The great number of loops was supposed to provide a more effective control of the net. In addition, gravity measurements were performed to reduce the observed height differences.

In 1966, the loss of a considerable number of bench marks made a re-levelling necessary. By the end of 1979, the data of about 80% of these measurements will be available. The missing 20% mainly represent subdivisions of large loops which support the existing net.

Determination of Levelling Accuracy

In general, the interval between two levellings varies between 10 years and several decades. Thus different instruments and methods are usually applied to the first and second measurement. Consequently, the causes of errors and their determination become of central interest for the investigation in vertical crustal movements.

The great variety of detailed studies shows how difficult it is to make any definite statement concerning the levelling accuracy. Many different methods have been applied so far: the least squares method, statistical analyses as well as the multivariant analysis (Stober, 1979; Hein, 1978). Nevertheless, none of these methods proves to be entirely satisfactory and apt to serve as the model for the determination of errors, as the errors are usually very complex. They have a number of sources: meteorological (temperature, atmospheric humidity) or topographical ones (refraction, position of turning plates) but there are also errors which depend on the equipment and the observer.

Statistical analyses could not be taken into account here, thus the calculation of the standard deviation had to be done according to the following formulas:

$$m_1 = \sqrt{\frac{1}{4n} \Sigma \frac{\Delta_1^2}{s}} \tag{1}$$

standard deviation for one km double levelling calculated from the height difference between direct and reciprocal levelling between two bench marks (Δ_1).

$$m_2 = \sqrt{\frac{1}{4n} \Sigma \frac{\Delta_2^2}{L}} \tag{2}$$

Standard deviation calculated from lines (Δ_2).

$$m_3 = \sqrt{\frac{1}{n} \Sigma \frac{\Delta_3^{2-}}{F}} \tag{3}$$

Standard deviation calculated from loop misclosures (Δ_3).

Table 1 comprises only a choice of the possible calculations of deviations. Nevertheless, it shows already considerable differences due to the individual computing methods. The calculation of accidental errors takes

Table 1

	m_1	m_2	m_3
Measurements 1948—1962	±0.54	±0.50	±1.94
Measurements 1966—1978	±0.35	±0.67	±0.72

an elimination of possibly existing systematical errors for granted. This, however, is not the case with m_1 and m_2. When calculating m_3 a great number of systematical errors (eg. subsidence of turning plates) have already disappeared. Curiously enough, m_3 is bigger than m_2 and m_1, especially for the period between 1948 and 1962. For the calculation of loop-misclosures the interval between the beginning and the end of the measurement is particularly important. A longer interval may already lead to remarkable height differences (>5 mm) at the point of departure (≡final point).

Table 2

Period (years)	1	2	3	4	5	6
Number of loops	4	6	9	6	3	4
m_3 (mm/km)	±2.16	±1.38	±1.84	±1.70	±1.80	±1.50

Table 2 shows us that there is no correlation between the length of time needed for carrying out these measurements and m_3. Some kind of erratic movement of the individual bench mark may be the cause of these results (eg. changing level of the ground water).

Adjustment of Those Measurements Carried out Between 1948—1962 and 1966—1978

Assuming that there is a constant velocity, the movement of the bench mark can be taken into account. Here the model developed by Hazay (1969) was used. According to this model the changes in height are calculated from two measurements. Then the levellings are reduced to the two different periods. In this case, results of the levellings performed between 1948 and 1962 were reduced to 1955, whereas those carried out between 1966 and 1978 were reduced to 1975.

The change per year is

$$\Delta_i = \frac{L_i'' + v_i'' - L_i' - v_i'}{dT_i}$$

where v_i are residuals and dT_i the period between two measurements. Furthermore

$$h_i' = L_i' + v_i' + \frac{t_i'}{dT_i}\,(L_i'' + v_i'' - L_i' - v_i')$$

are adjusted, reduced height differences (first period) and

$$h_i'' = L_i'' + v_i'' + \frac{t_i''}{dT_i}\,(L_i'' + v_i'' - L_i' - v_i')$$

are the same for the second period, where t_i is the time between the actual measurement and the year to which it was reduced and C is a correction by means of gravity measurements.

The adjustment was carried out under the following conditions: $[h'] - C = 0$ and $[h''] - C = 0$. The reduced and adjusted height differences resulting from these calculations can easily be converted into vertical velocities. An arbitrarily chosen bench mark, whose vertical movement is ignored, serves as a reference.

The levellings used for this adjustment do not comprise the western part of Austria. There are no real possibilities of forming loops in that part of the Austrian territory. Within one period the individual measurements are considered to be of equal weight. The two periods were weighted according to m_1 (Table 1):

$$\frac{m_1^{1955}}{m_1^{1975}} = \sqrt{\frac{w_{1975}}{w_{1955}}} \qquad w_{1955} = 1 \qquad w_{1975} = 2.5$$

As a result of the adjustment the following standard deviations are to be found:

$$m_h = \pm 11.2 \text{ mm} \qquad \text{(period 1955)}$$
$$m_h = \pm\ 7.1 \text{ mm} \qquad \text{(period 1975)}$$

If we suppose the distance between two bench marks to be 50 km, we get the following standard deviation of 1 km

$$m_0 = \pm 1.6 \text{ mm/km} \qquad \text{(period 1955)}$$

$$m_0 = \pm 1.0 \text{ mm/km} \qquad \text{(period 1975)}$$

and for the difference between two measurements we get

$$m_h = \pm 1.9 \text{ mm/km}$$

reduced to one year

$$m_v = \pm 0.1 \ \sqrt{s \ [\text{km}]} \quad \text{(mm/year)}$$

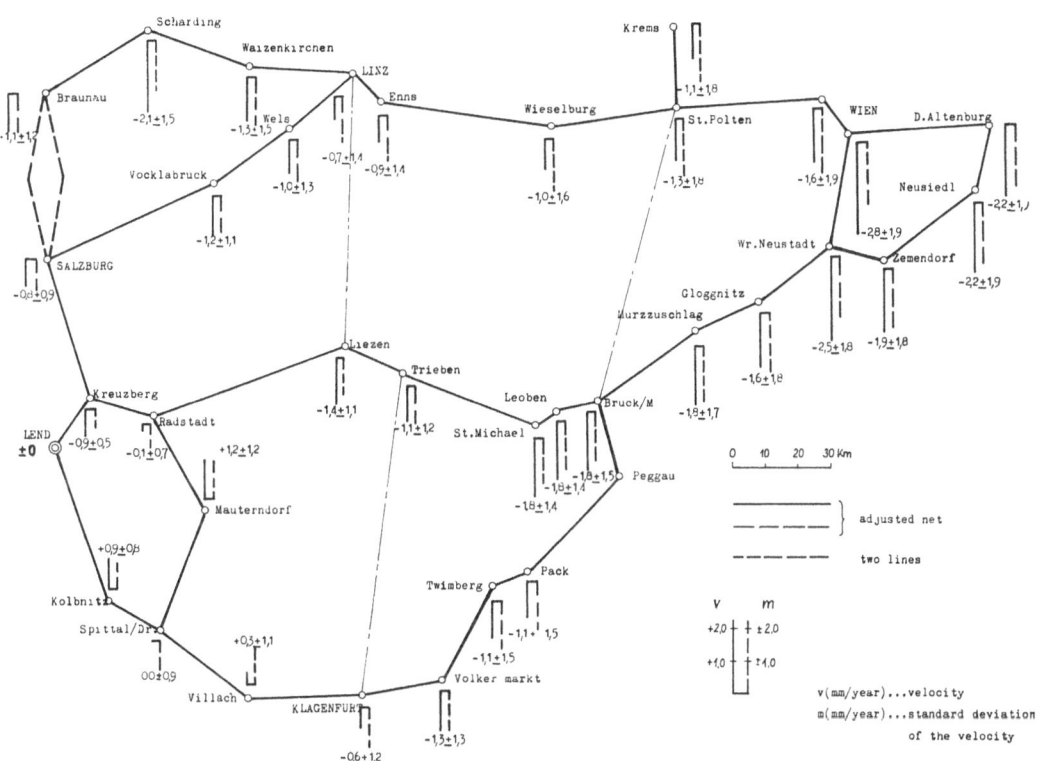

Fig. 1. Vertical velocities in Austria

Results

A bench mark group near Lend serves as a reference here. These marks are situated in the rocks of the Alps and thus not so much subject to limited regional changes. That is why they were chosen as reference.

A general analysis given in Table 3 shows that — with the only exception of Schärding — no significant movement can be found near the

Bohemian Massif. These subsidence velocities may — at least partly — be due the fact that the bench marks available in this region are only to be found on buildings. Apart from the individual vertical movements, an uplift of the bench marks at Lend seems to be likely. On the whole, drawing any

Table 3

	v (mm/y)	m_0 (mm/y)	S (%)
Lend	0	—	—
Salzburg	−0.8	±0.9	63
Braunau	−1.1	±1.2	63
Schärding	−2.1	±1.5	84
Waizenkirchen	−1.3	±1.5	60
Linz	−0.7	±1.4	< 40

definite conclusions with regard to the Alps and their relationship to the Bohemian Massif is questionable. The main reason for this is the fact that no direct levelling line running from the Bohemian Massif to the Alps has been re-measured yet.

Within the Alpine region, vertical movements are to be found with the reference to the bench marks in the valleys. When dealing with these movements, however, we must also keep in mind that the differences in altitude are the source for additional errors. These errors appear during the comparison of the rod on the one hand and on the other hand during the actual measurement. Thus we must take into account up to ±0.04 mm/m difference of altitude. Accordingly, the calculated movements are of less significance than those computed from the adjustment.

Table 4

	ΔH (m)	δ_H (mm)	$m_0{}^a$	$m_0{}^c$	S_a (%)	S_c (%)
Gloggnitz/ Semmering	545	−6.9	±7.1	±22	68	< 40
Twimberg/Pack	577	+1.5	±8.3	±23	< 40	< 40
St. Michael/St. Wald/ Schoberpaß	170	+12.0	±11.0	±7	73	98
Mauterndorf/ Obertauern	630	+14.0	±7.8	±24	93	58

$m_0{}^a$ (mm) ... Standard deviation calculated from the adjustment.
$m_0{}^c$ (mm) ... Standard deviation calculated with due regard to the rod error.

In general, we must say that the velocities of the bench marks available at the moment include a number of errors, as there is only a very short period between the two measurements compared. Thus no definite statement with regard to uplifts and subsidences can be made at the moment from our viewpoint.

The eastern part of Austria forms the only exception in this connection. Therefore the Wiener Becken will be treated in greater detail.

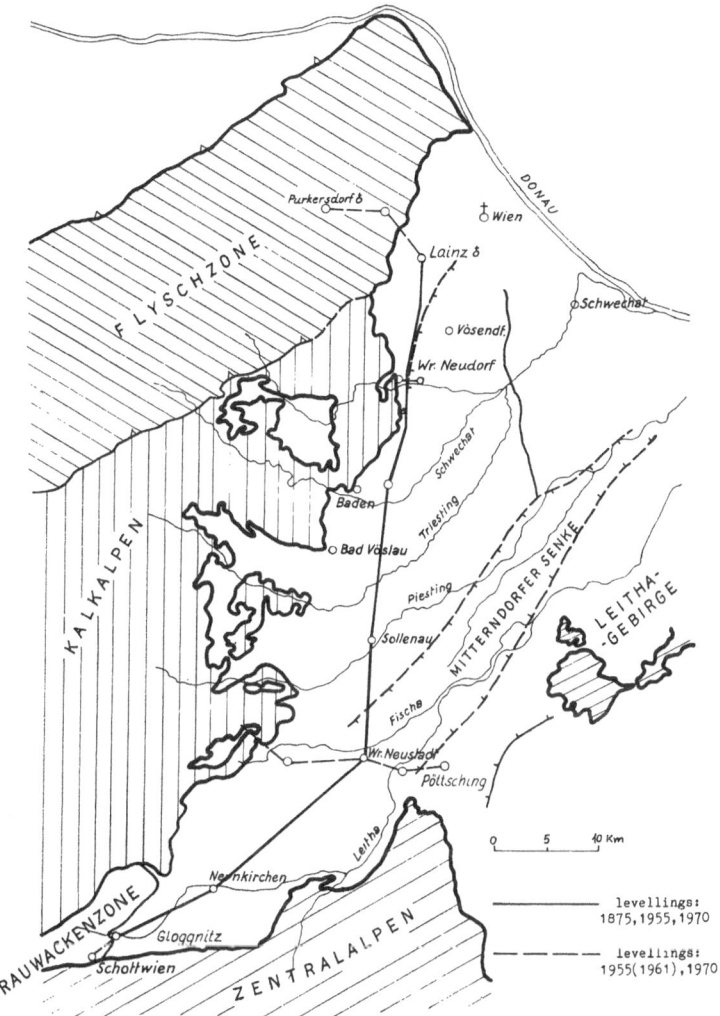

Fig. 2. General map of the geological zones of the Wiener Becken

Investigations in the Wiener Becken

As Fig. 1 already shows, the Wiener Becken produces the maximum velocities of bench marks within the Austrian territory. Thus the particular interest in this region. Further, the existence of a number of side levellings beside the main levelling line, which crosses the area, has made it possible to compare the results and to get some substantial evidence.

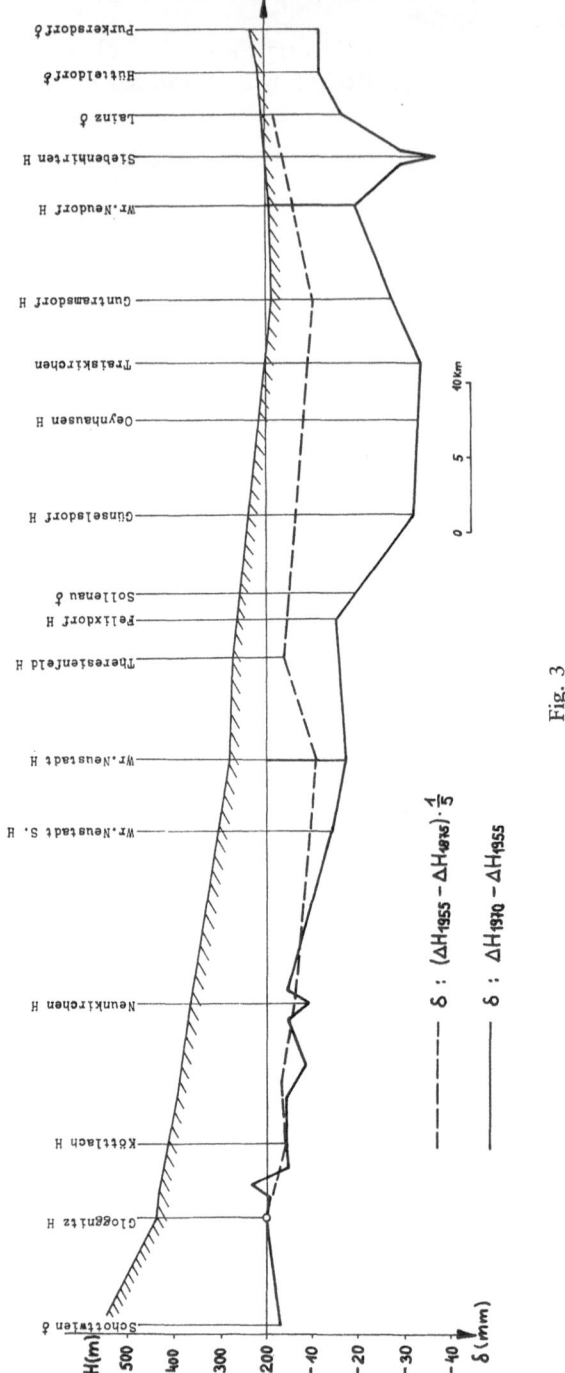

Fig. 3

According to Fig. 2, the levelling line parts from Purkersdorf, to the east of Vienna, in order to reach the Wiener Becken soon after Hütteldorf and then runs southward (following the federal highway 17), passes Wr. Neustadt and then follows the road to Schottwien, a village to the southwest of Wr. Neustadt.

The side levellings run from Mödling to Biedermannsdorf via Wr. Neudorf as well as from the Hohe Wand to Pöttsching via Wr. Neustadt.

For this investigation, levelling results from three different measuring periods were used. The oldest ones date back to period 1875 (1873—1877). At that time the line between Gloggnitz and Lainz was measured. During the periods 1955 and 1970 re-levellings were performed. Only the side levelling from the Hohe Wand to Wr. Neustadt was done in 1961 and 1972.

Comparison Between the Levelling Results of Period 1875 and 1955
(Fig. 3)

Here the actual results of the levelling and not the adjusted ones were used. For the comparison of these two measuring periods only 11 of the old

Table 5

	m_0 (mm)	δ_H (mm)	S (%)
Gloggnitz	—	0	—
Neunkirchen	±18.0	−33	93
Wr. Neustadt, St.	±25.6	−56	97
Lainz	—	0	—
Biedermannsdorf	±11.5	−21	93
Guntramsdorf	±16.1	−42	99
Wr. Neustadt, St.	±30.3	−56	93

$m_0^{1875} = \pm 4.5$ mm/km

$m_0^{1955} = \pm 1.4$ mm/km

bench marks were available. The little number of bench marks does not seem to be sufficient to draw any definite conclusions. There may be changes which concern only one single mark which cannot, however, be separated from the general tendency.

Comparison of the Levelling Results from Period 1955 and 1970
(Fig. 3)

Due to the great number of bench marks (about 100), the instability of the individual bench marks could be excluded. Here again the bench mark at Gloggnitz serves as a reference.

After having reduced the results of 1875—1955 to a period of 15 years, a similar tendency was to be found as in those received between 1955 and

Table 6

	m_0 (mm)	δ_H (mm)	S (%)
Schottwien	±3.9	−2.9	54
Gloggnitz	—	0	—
Wr. Neustadt, J.	±8.6	−17.4	95
Guntramsdorf	±12.2	−34.1	99
Lainz	±13.4	−16.7	77
Purkersdorf	±14.0	−11.5	58

1970. An increase of the subsidence velocities in most parts of the Wiener Becken could be observed.

Side Levellings

Fig. 5 shows us a diagram of levelling results from the line between Mödling and Biedermannsdorf. With reference to the bench marks at Hütteldorf and Purkersdorf, a subsidence of the bench mark at Wr. Neudorf of 8 mm within 15 years ($v = -0.5$ mm/y) must be stated. With regard to the

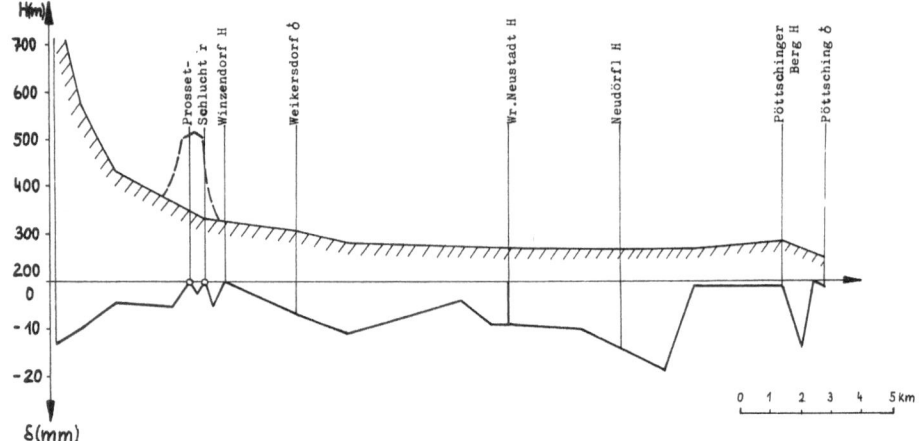

Fig. 4

bench mark at Mödling, however, there is a subsidence of 13 mm within 22 years ($v = -0.6$ mm/y). These results are indeed in good agreement. The subsidence in this region seems to be particularly interesting as it appears within 1 km between Mödling and Wr. Neudorf.

The side levelling at Wr. Neustadt is far more extensive: it runs from the Hohe Wand to Wr. Neustadt and then goes on to Pöttsching. As point of departure, two bench marks placed in the rocks of the Fischauer Vorberge were chosen. The junction point at Wr. Neustadt is representative of its surroundings and could be used for a comparison between both the main

Fig. 5

line and side levellings. With regard to the Fischauer Vorberge a subsidence of 10 mm within 11 years (-0.9 mm/y) could be observed, whereas with regard to the marks at Schottwien 15 mm within 15 years (-1.0 mm/y) were to be found. These two measurements are in good agreement too. Remarkable again, the course of the line east of Neudörfl: within 1 km there is an uplift of 15 mm within 17 years and then reaches approximately the same level as the bench marks in the Fischauer Vorberge.

Conclusion

The levellings that have been done in the Wiener Becken so far, are in good agreement and show a subsidence which confirms the geological and geophysical investigations.

When dealing with more spacious movements, however, such as those concerning the area between the Alps and the Bohemian Massif, the results still include a number of errors. On the one hand, the reasons for these errors are to be found in the differences in altitude, on the other hand, the short period of time which lies between the individual measurements seems to exercise a certain influence. According to the small interval the changes in height are less significant.

On the whole, we can say that the levelling results make certain tendencies evident. Yet, only future investigations can furnish proof.

References

Ghitau, D.: Modellbildung und Rechenpraxis bei der nivellitischen Bestimmung säkularer Landhebungen. Dissertation, Bonn (1970).

Hazay, I.: Die Ausgleichung von Nivellementsnetzen für die Beobachtung der vertikalen Erdkrustenbewegungen. Reprint from the Publications of the Technical University for Heavy Industry, Miskolc, Hungary 29 (1969).

Hein, G.: Multivariate Analyse der Nivellementsdaten im Oberrheingraben und Rheinischen Schild, ZfV 1978 (9) (1978).

Holdahl, S. R.: Models and Strategies for Computing Vertical Crustal Movements in the United States. Int. Symposium on Recent Crustal Movements, Grenoble (1975).

Schlemmer, H.: Laser-Interferenzkomparator zur Prüfung von Präzisionsnivellierlatten. DGK, Reihe C, Heft 210 (1975).

Stober, M.: Zur Erfassung rezenter vertikaler Krustenbewegungen durch Präzisionsnivellements. Dissertation, München (1979).

Address of author: Dipl.-Ing. N. Höggerl, Bundesamt f. Eich- u. Vermessungswesen, Abt. K2, Friedrich-Schmidt-Platz 3, A-1080 Wien, Austria.

Rock Mechanics, Suppl. 9, 213—217 (1980)

Rock Mechanics
Felsmechanik
Mécanique des Roches
© by Springer-Verlag 1980

Theme 6
Geomechanical Models

On the Nature of the Alpine Stress Field

By

Horst J. Neugebauer

With 2 Figures

Abstract

The present stress field of the Swiss Alps, as it becomes evident from various observational methods, is discussed with regard to its specific character. In order to understand the nature of the observations, different mechanisms of crustal loading will be considered.

By means of the results of two-dimensional numerical experiments on the dynamics of the Alpine crustal structure, the influence of particular load systems on the modelled stress field will be specified.

These considerations strongly support the view that the present stress field of the Alps is dominated by the structural and morphological features of the crust.

The Stress Field

Seismotectonic, tectonic and joint-investigations in addition to in-situ measurements reveal information on the crustal stress field. Both, focal mechanism studies as well as in-situ measurements exhibit the NNW—SSE orientation of the maximum horizontal compression as a major trend of the stress field (Pavoni, 1976; Greiner and Illies, 1977; Schneider, 1979; Froidevaux et al., 1979). This coincidence is disturbed at such tectonic active areas as the Rhenish Massif or the Alpine mountain belt. While for the Rhenish Massif the orientation of the maximum horizontal compression turns from NNW to NW, a change of orientation towards WNW can be recognized for the Western Alps, associated with the overall strike of the mountain belt (Pavoni, 1976, 1979; Ahorner, 1978; Neugebauer and Tobias, 1977). Thus it becomes evident from the focal data that the orientation of the crustal stress field might be affected by mountain areas.

A similar situation can be stated for the amount of crustal stresses on the base of in-situ determinations. Froidevaux et al. (1979) found maximum horizontal stresses to be typically 20 bar (2 MPa) at the Parisian and Acquitaine Basins. This mean level will be met by measurements presented by

0080-3375/80/Suppl. 9/0213/$ 01.00

Greiner and Illies (1977) for the northern Alpine foreland. On the contrary, the excess of horizontal stress reveals maximum values of more than 200 bar (20 MPa) at the Alpine fold belt. Hence, these data display again a possible Alpine influence on the stress field of the crust.

Crustal seismicity can be understood as an indicator of the present stress field. Although it is rather difficult to read historical and recent seismicity in terms of features of the stress field, an outstanding conclusion concerning the Alpine situation should be emphasized here. Pavoni and Mayer-Rosa (1978) associated a zone of relatively high seismicity along the strike of the Alps on the northern border of the autochthonous massifs with the Cenozoic hinge zone between Alps and Molasse foredeep. This consideration supports the view that the specific spacial concentration of strong earthquakes is likely to indicate corresponding features of the stress field.

On the other hand, the Swiss Alps are supposed to be in an active phase of recent crustal movements which has been related to the present state of isostatic compensation (Kahle et al., 1976; Neugebauer et al., 1980).

Numerical Models

In order to reach a better understanding of the principal features of the stress field associated with a structure like that at the Alpine mountain belt, numerical models could be a powerful tool. The numerical calculations which will be cited are based on two-dimensional, plane strain finite element

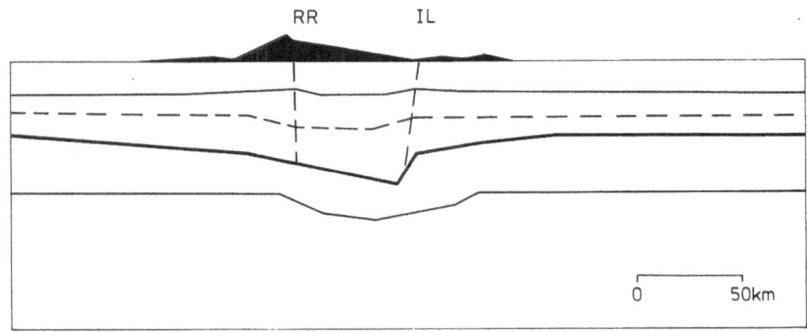

Fig. 1. Modelled cross section through the Swiss Alps derived from Müller et al., 1977

models with nonlinear creep rheology. The adopted structural features are the result of extensive geophysical studies along the Swiss Geotraverse: Basel — Chiasso, Fig. 1. According to the model structure of the crustal cross section, the applied load system is composed by the excess of body forces with respect of a "normal crust" according to the mountain relief and the buoyancy corresponding to an Airy-Heiskanan density model of the crust and mantle after Müller et al. (1977).

The applied concept of isostasy provides a negative isostatic anomaly which has been interpreted as an additional buoyancy force exceeding the required amount for isostatic equilibrium. In addition, external horizontally-acting tectonic forces have been considered due to the concept of plate tectonics and its implications for the Alps. An extensive discussion of the whole sequence of models is given by Neugebauer and Brötz (1979).

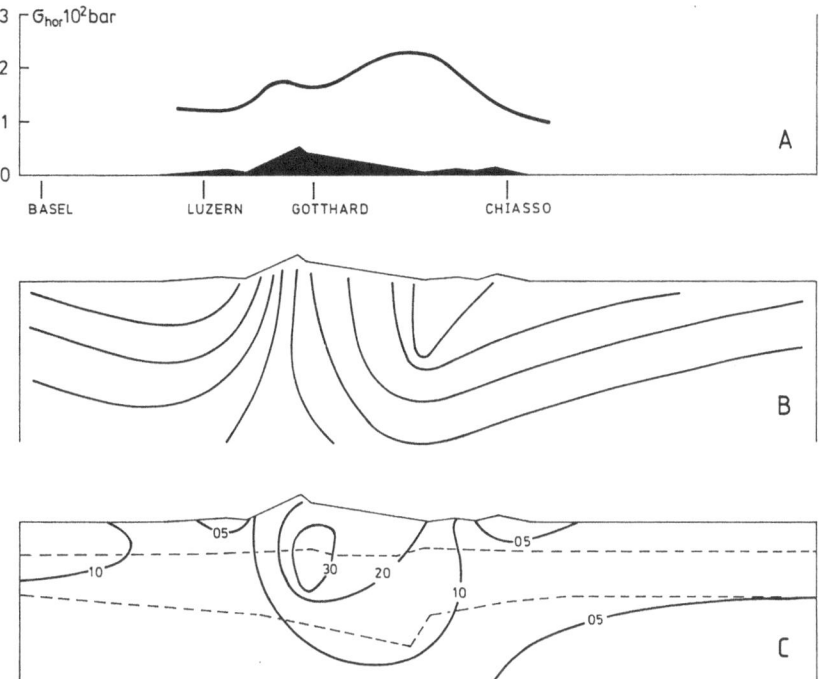

Fig. 2. Horizontal stress component, orientation of maximum principal stress and distribution of maximum shear stress in 10 bar from above to below

Fig. 2 summarizes the main features of the modelled stress field. In order to represent the behaviour of a crustal plate, the crustal boundary at the left and the right at the shown section, Fig. 1, has been fixed. In the context of the applied body-forces, the fixed boundary acts in the horizontal direction like a push of about 250 bar or 2.5×10^{10} N/m respectively. Thus, the presented stresses are induced by the composition of structural and applied external loads.

Discussion

The calculated horizontal stress component, Fig. 2 A, of the above model shows its maximum within the zone of maximum crustal thickness which on the other side coincides with the maximum of recent crustal uplift. The deviation of the maximum amount of the horizontal stress component from the high mountain relief is the result of the density model chosen and the corresponding excess of buoyancy associated with the isostatic anomaly. The flanks of the curve will decrease more rapidly towards a low stress level for an appropriate lower horizontal external force. Thereafter a pronounced high of horizontal stress components at the Alpine model structure is induced by the differential loads with respect to the crustal structure. It will furthermore be affected by the existence of excess buoyancy in the case of an isostatic nonequilibrium. So the modelled parameter suggest that the maximum is not necessarily the consequence of high horizontal external or tectonic loads (Neugebauer et al., 1980).

Fig. 2 B shows the orientation of the calculated maximum principal stresses. The indicated dip of the trajectories towards the mountain section and their turn up within this section is a characteristic result which will be lost in models with higher horizontal loads than the applied one. According to the modelled features, we would observe a predominantly horizontal compression, however, with modified dip.

The corresponding maximum shear stress contours, Fig. 2C, exhibit a maximum at the crustal structure associated with high-relief mountains. The asymmetric decline of the shear stress maximum as well as its relative amount becomes more distinctly marked for declining horizontal external forces. With respect to the seismicity even this field parameter would point towards a relatively low tectonic contribution to the stress field.

According to the presented parameters of the modelled stress field of the Alpine crustal structure, one will recognize a characteristic "Alpine contribution" to the stress field in contrast to the "foreland" for a rather low horizontal tectonic force. The expressed character will on the contrary be governed by the structural and isostatic conditions of the crust.

References

Ahorner, L.: Untersuchung von Mikroerdbeben im Bereich des Rheinischen Schildes. Protokoll über das 3. Kolloquium im Schwerpunkt „Vertikalbewegungen und ihre Ursachen am Beispiel des Rheinischen Schildes", Bad Godesberg (1978).

Froidevaux, C., Paquin, C., Souriau, M.: Tectonic Stresses in France. Trans. Am. Geophys. Union 60, 607 (1979).

Greiner, G., Illies, H.: Spannungsmessungen: Probleme, Methoden, Ergebnisse. Jahresbericht 1976, Sonderforschungsbereich 77, Felsmechanik, Universität Karlsruhe (1977).

Kahle, H.-G., Klingele, E., Müller, St., Egloff, R.: The Variation of Crustal Thickness Across the Swiss Alps Based on Gravity and Explosion Seismic Data. Pageoph 114, 479—494 (1976).

Müller, St., Kahle, H.-G., Kissling, E.: Seismik und Schwere entlang der Schweizer Geotraverse. Geodynamics and Geotraverses Around the Alps, Salzburg (1977).

Neugebauer, H. J., Brötz, R.: The Present Stress Field of the Swiss Alps and its Possible Origin. Submitted to Pageoph (1979).

Neugebauer, H. J., Brötz, R., Rybach, L.: Recent Crustal Uplift and the Stress Field of the Alps Along the Swiss Traverse Basel-Chiasso. In press. Eclogae geol. Helv. (1980).

Neugebauer, H. J., Tobias, E.: A Study of the Echzell/Wetterau Earthquake of November 4, 1975. J. Geophys. *43*, 751—760 (1977).

Pavoni, M.: Herdmechanismen von Erdbeben und regionaltektonisches Spannungsfeld im Bereich der Geotraverse Basel – Chiasso. Schweiz. mineral. petrogr. Mittl. *56*, 697—702 (1976).

Pavoni, M.: Crustal Stresses Inferred from Fault-Plane Solutions of Earthquakes and Neotectonic Deformation in Switzerland. Trans. Am. Geophys. Union *60*, 607 (1979).

Pavoni, M., Mayer-Rosa, D.: Seismotektonische Karte der Schweiz 1 : 750000. Eclogae geol. Helv. *71*, 293—295 (1978).

Schneider, G.: Seismic Stresses in Southern Germany. Trans. Am. Geophys. Union *60*, 607 (1979).

Address of author: H. J. Neugebauer, Institut für Geophysik, Technische Universität Clausthal, Arnold-Sommerfeld-Strasse 1, D-3392 Clausthal-Zellerfeld, Federal Republic of Germany.

Rock Mechanics, Suppl. 9, 219—232 (1980)

Rock Mechanics
Felsmechanik
Mécanique des Roches
© by Springer-Verlag 1980

Stress Distribution
in Overthrusting Slabs and Mechanics
of Jura Deformation*

By

W. H. Müller and K. J. Hsü

With 6 Figures

Abstract

Our investigations relate the stability to stress-distribution in overthrusting slabs. Finite-element models have been formulated to evaluate the magnitude and orientation of stress within overthrusting slabs under given boundary conditions of displacement. The Mohr-Coulomb criterion has been adopted to define a zone of failure. In addition to a simple model to describe idealized slabs, the tectonic deformation of the Jura decollement has been modelled.

The computations were carried out with a program which enables the simulation of elastic/ideal plastic behavior. For the first simulation the stability of an overthrusting crustal slab is considered, assuming a left-to-right displacement with the right hand edge fixed expect for the lowest 1 km thick layer. The cases of a completely homogeneous slab, of a slab with one layer of lower strength, and of a slab with two low-strength layers, were considered in different simulation experiments. Deformation of the homogeneous body could not produce a clearly defined detachment-surface. In the one-layered block, the low-strength layer acts as a detachment-surface. The ratio of the length of the thrust plate to its thickness is 4/1 (16 km long for a 4 km thick thrust, or 40 km long for a 10 km thick plate) in this test. However, by varying the composition of the layering, the length/thickness ratio of a multiple thrust complex can be 20/1 or 80 km long for a 4 km thickness plate. This ratio is comparable to that observed in the cover nappes and rigid-basement nappe-complexes of the Alps.

In the case of the Jura-overthrust, two different models have been investigated: by rotation on a concave detachment horizon ("distant-push" theory) and subduction of the crystalline basement under the Aare-massif ("underthrusting" theory). As ductile shear-horizon, we assumed a 100-m thick layer of anhydrite. The material constants have been determined by triaxial deformation experiments. The analyses clearly show tha the Jura mountains could hardly have been deformed in the manner postulated by the "distant-push" theory. On the other hand, the assumption of the underthrusting model can result in satisfactory simulation of field observations.

---

* Contribution No. 145 of the Laboratory of Experimental Geology.

0080-3375/80/Suppl. 9/0219/$ 02.80

Introduction

The mechanical difficulty of moving large bodies of rocks such as Alpine overthrusts was first raised by Smoluchoski (1909); the friction at the base should be so great as to prevent their transport. Hubbert and Rubey (1959) presented a lengthy review and offered a suggestion that the friction could be reduced because of the presence of abnormally high pore-pressure in the zone of thrusting. Hsü (1969) pointed out that the thrusting may have taken place along a thin zone of plastic deformation and that the overthrusting mechanics cannot be treated simply as a problem of frictional sliding. However, all those authors treated the problem as one of mechanical transport along a pre-existing or a predestined horizon; the question of initiating a decoupling horizon or of failure along a plane of thrusting was bypassed.

Anderson (1942) reasoned that reverse faulting in an isotropic and homogeneous material under homogeneous stress should take place along a plane making a 30 degree angle with the direction of principal compression. Hafner (1951) and Couples (1977) treated the question of localization of failure for isotropic and homogeneous medium under heterogeneous stress; according to their analysis, the body of a rigid thrust-plate should be the "stable region", where the induced shear stresses are smaller than the critical shear stress τ_c as given by the Mohr-Coulomb criterion:

$$\tau_c = \tau_0 + \sigma \tan \varphi_c. \tag{1}$$

In this equation, τ_0 is cohesion, σ_n the normal stress, and φ_c the angle of internal friction. Within the "unstable region" or zone of failure, the shear fracture should take place along a surface making a 30 degree angle with the direction of principal compression. If the surface separating his regions of stability and of failure is a thrust plane, Hafner's analyses showed that this thrust plane cannot have a great horizontal extent. However, he did not question why large overthrusts tended to move along horizontal planes.

Geological observations suggested that flat-thrusts are possible because they took place along "zones of weakness". The existence of such zones implies the heterogeneity of the medium under stress. Analytical methods prior to 1960 were not yet sufficiently advanced to handle the question of heterogeneous medium under heterogeneous stress. With the innovation of the finite-element method, we were finally ready to tackle this question.

This article reports our study of the stress-distribution within a layered medium under various boundary stresses induced by variable displacements. We demonstrated that flat-thrusts are indeed possible and that detachment or decoupling could take place along zones of reduced strength. We further analysed the tectonics of the Jura Mountains and found that underthrusting is the most probable mechanism to induce the Jura decollement.

Methodology

The principle of using the finite-element method to analyse a stressed medium is well known (e. g. Zienkiewicz and Cheung, 1967) and the application of this method to geology has been discussed in several previous publications (e. g. Dietrich and Carter, 1969; Dietrich and Onat, 1969; Voight and Samuelson, 1969; and Bock, 1972).

For our work we used a finite-element computer program STAUB (Statische Analyse von Untertags-Bauten) developed by the Civil Engineering Department of ETH, Zurich (Kovari, 1969) based upon the procedures described in Zienkiewicz and others (1967, 1969). The STAUB program assumes an elastic-ideally plastic behavior of the stress medium, while a part of the system is elastically strained, another part may be plastically deformed. In the plastic zone the strain includes an elastic and a plastic increment. Local plastic zones are bounded by zones which remain elastic and the boundary conditions hinder an unlimited plastic flow. This limitation permits the program to function until stress at all points exceeds the yield criterion.

The STAUB program utilizes the yield criterion of Drucker and Prager (1952), namely

$$f(I_1, J_2) = I_1 - J_2 + K = 0 \qquad (2)$$

where

$$I_1 = \sigma_1 + \sigma_2 + \sigma_3$$

$$J_2 = \frac{1}{6}(\sigma_1 - \sigma_2)^2 + (\sigma_2 - \sigma_3)^2 + (\sigma_3 - \sigma_1)^2$$

$\sigma_1, \sigma_2, \sigma_3$ being principal stresses, and k defined by

$$K = \frac{3\,\tau_0}{\sqrt{9 + 12\tan^2\varphi_c}}$$

In the case of two-dimensional plastic flows, it has been shown that this yield criterion is equivalent to the Mohr-Coulomb-Navier criterion, which has the same form as the Mohr-Coulomb criterion for fracture (Jaeger, 1969). In other words, Eq. (1) is valid as an expression of fracture-criterion for the brittle, and of the yield stress, for the ductile deformation (Hsü, 1969; Handin, 1969).

Homogeneous Slabs Under Stress

Assume a homogeneous slab 5 km thick and 50 km long and divide the slab into 250 triangular elements. Push the slab from left to right (Fig. 1). All the points on the left side, on the base, and the lowest two points on the right side could be displaced. The upper 3 points are on the right hand and were held in a fixed position. Assume further that the points on the base can only be displaced in a horizontal direction, whereas the points on the upper surface have the freedom of movement. We could thus simulate an underthrust from the left, or an overthrust from the right. The mechanical property of the slab is given as follows: Young's modulus $E = 600$ kilobars;

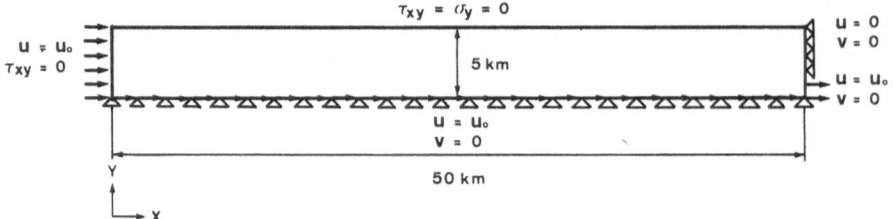

Fig. 1. Dimension and boundary conditions of the slabs in simulation experiments

Poisson's ratios $v = 0.3$; Bulk density $\varrho = 2.6$ g/cm^3; Angle of internal friction $\varphi_c = 30^0$; and Cohesion $\tau_0 = 300$ bars.

Fig. 2 shows the development of the region of instability within the thrust plate after 10-m, 150-m and 300-m displacement of the slab. There is no clear-cut horizon of decoupling within this homogeneous slab to permit the genesis of flat thrusts. The stress-distribution is also shown. As one

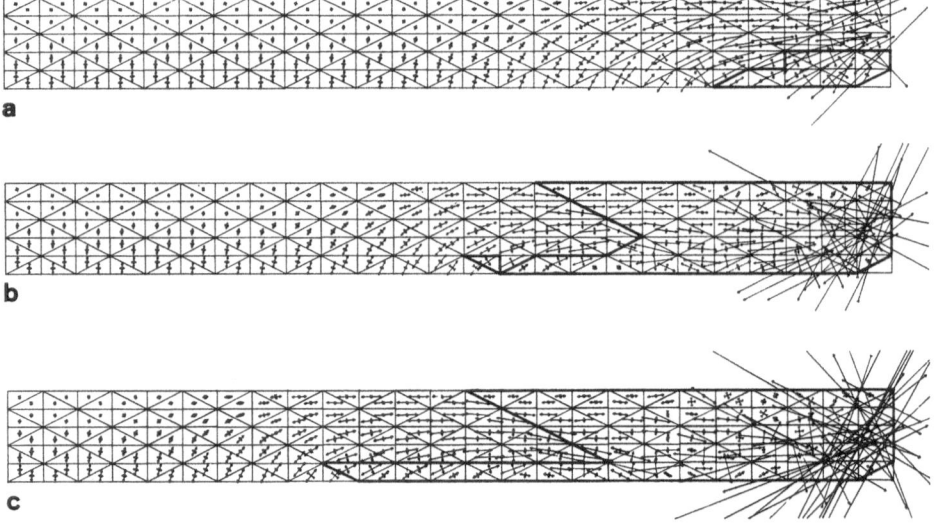

Fig. 2. Stress distribution of a homogeneous slab being displaced by a push of (a) 10 m, (b) 150 m and (c) 300 m from behind. Unstable regions where the yield criterion has been satisfied are delimited by heavy lines. The arrows show the direction of principal stresses and the length of arrows gives a measure of magnitude (3.1 mm = 2 kb). Note the prevailing horizontal compression near the right end and the extension near the left end

expects, the orientation of the principal stress is largely horizontal; compression should occur near the right side and thrust faulting where the plate is being shortened (Fig. 2). However, we were surprised to find that the maximum principal stress (σ_1) is nearly vertical near the left side, where gravitational stress is dominant. The lesser horizontal compression in this region is a Poisson's ratio effect and can be described in this large self-gravitating body by

$$\sigma_{\text{vert}} = g \cdot z \cdot \varrho$$

$$\sigma_{\text{horiz}} = \frac{v}{1-v} \cdot \sigma_{\text{vert}}$$

where ϱ = density, g = acceleration due to gravity and v = Poisson's ratio. In such a stress system, we do not have a hydrostatic state but σ vert > σ horiz unless $v = 0.5$. Whether the "standard state" (Hafner, 1951) corresponds to real nature for the stress situation, or whether we have to assume Heim's rule — that the natural state of stress tends to become hydrostatic over a long period of time (Jaeger, 1969, p. 172) — is still a subject of great discussion (Couples, 1977), and we are still not able to answer this question.

In several series of computer-modelling with a homogeneous medium, we changed the values of the various parameters of the displacements, and of the mechanical properties and obtained similar stress patterns. In no case did we find a flat thrust moving along a horizontal zone of decoupling (Müller, 1975).

Layered Slab Under Stress

We assume the same slab geometry and boundary condition as in the previous test and the same values for the mechanical properties, except that a layer at 3—4 km depth has a cohesion of 15 bars, or 1/20th that of the

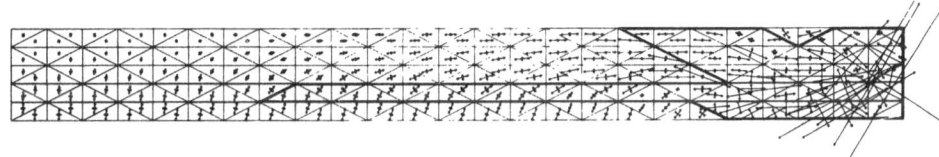

Fig. 3. Stress distribution of a layered medium being displaced by push of 50 m from behind. Note the development of a yield horizon along the layer of low cohesive strength. (See text for detailed explanation)

surrounding rocks. Fig. 3 shows the region of instability within the thrust plate after a 50-m displacement. As one expects, this layer of low cohesive strength becomes a horizon of decoupling, permitting a relative horizontal displacement of the rigid slabs above and bleow. This computer-simulation provides the mathematical confirmation that the Alpine type of overthrusting is possible only if there have been a zone (or zones) of weakness to permit failure or yielding along this decoupling-horizon. The ratio of the length of the "rigid point" of the thrust plate to its thickness is 4/1 in the test shown by Fig. 4 (16 km long for a 4 km thick thrust, or 40 km long for a 10 km thick plate). However, this ratio can be considerably larger by varying the various mechanical parameters, or by varying the composition of the layering. Fig. 4 shows, for example, that the length-thickness ratio of a multiple

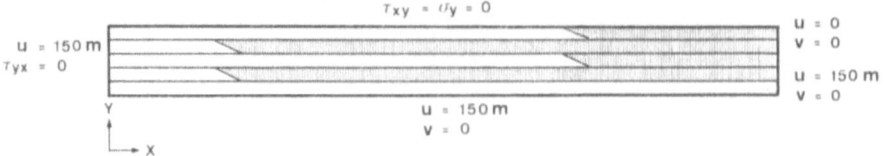

Fig. 4. Development of thin and long thrust plates in a layered medium with more than one layer of low cohesive strength. (Inclined hatched area shows regions of instability)

thrust-complex can be 20/1 or 80 km long for a plate 4 km thick. This ratio is comparable to that observed in the cover nappes and rigid-basement nappe-complexes of the Alps.

Geological Significance

The original project was to study the stress distribution within thrust-plates prior to their displacement, involving mainly sedimentary formations. Such overthrusts in the Helvetic Alps include mainly sedimentary formations. The analysis provides a quantitative demonstration of the well-known geological observation that decollement-type of so-called "cover-thrusts" tend to form along zones of weakness, such as flysch, anhydrite or shalyformations (Trümpy, 1969). The question may be asked why large rigid-basement thrusts, such as the Silvretta Nappes of the Austro-Alpine Nappe-Complex are possible; the thrustplane or the detachment surface cuts across a massive, and apparently homogeneous granitic or metamorphic massif (see Trümpy, 1960). Our studies on the structure of and on the temperature-distribution within the crust under the Alps show that the strength within a materially homogeneous massif may not be uniform (Finckh, 1976). The low-velocity layer in the crust where the rock has a temperature near or at partial melting could have considerably less strength by an order of magnitude or so, than that of the overlying slab. Such a low-velocity layer could be the decoupling horizon underlying rigid-basement nappes (or crustal) plates.

Mechanics of Jura Deformation

There exist two different hypotheses about the origin of the Jura mountains. One school of thought, also known as the distant-push (Fernschub) theory, suggests no crustal shortening between the Aare-Massif and the Table Jura: the shortening of the Jura cover was induced by the northward advance of the Helvetic and Molasse rocks above the concave Triassic detachment surface, and that advance in turn was related to rotational gravity-sliding (model 1, variant A), aided perhaps by a distant push in the region of the Aare-Massif (model 1, variant B) (Buxtorf, 1907; Laubscher, 1961). Another school of thought was to postulate an underthrusting of the Jura-basement under the Aare Massif (model 2) (Umbgrove, 1948; Hsü, 1979). Both hypotheses assumed the presence of a very weak layer in the decollement horizon, which was deformed by ductile gliding.

To determine the stress response to boundary conditions of displacement, one must define the gliding horizon. Experimental investigations indicate that anhydrite is more ductile than limestone, and that fine-grained anhydrite is more ductile than coarse anhydrite during creep tests at very small strain rates (Müller and Briegel, 1977, 1978). Examination of drill cores from the detachment horizon (Keuper, Altishofen, at 2000 m depth, see Fischer and Luterbacher, 1963) indicated that the fine-grained anhydrite has indeed undergone far more intensive ductile deformation than the coarser anhydrite and the overlying limestones. We concluded, therefore, that the fine-grained anhydrite from the Keuper and from the Anhydrite group of the Triassic is ductile gliding material during the Jura deformation (Müller and Briegel, in prep.).

To make simulation-computations, we assume the Mohr-Coulomb criterion of failure for rocks outside of the gliding horizon, and an empirical flow-law for plastic yielding of the anhydrite within the gliding horizon. Where the stress difference is less than the critical shear stress or the creep strength, the response is elastic, where the stress difference is greater, the deformation of the element is plastic.

For stress-computations outside the gliding zone we need data of the material properties of the crystalline basement, of the limestone and of anhydrite; the following assumptions were made on the basis of available experimental data (see Clarke, 1966; Müller and Briegel, 1977):

		Crystalline	Anhydrite	Limestone
Young's modulus	$E =$	0.8 mbar	0.25 mbar	0.6 mbar
Poisson's ratio	$\nu =$	0.3	0.3	0.3
Density	$\varrho =$	2.6	2.9	2.6 g/cm³
Cohesive strength	$\tau_0 =$	800	400	970 bar
Coefficient of internal friction	$\varphi_c =$	40	30	24

To determine the yield strength of the anhydrite within the gliding horizon, we need information on the rheological behavior of such an anhydrite. The flow-law experimentally determined by creep tests, is according to Müller and Briegel (in prep.):

$$\dot{\varepsilon} = A \cdot \exp \cdot \left(\frac{-H}{R \cdot T} \right) \cdot \left(\sinh \frac{\sigma}{\sigma_0} \right)^n$$

where $\dot{\varepsilon}$ is strain rate; A, n and σ_0 are constants; H is the apparent activation energy for creep; R is the gas constant; and T is the temperature in ⁰K. Four

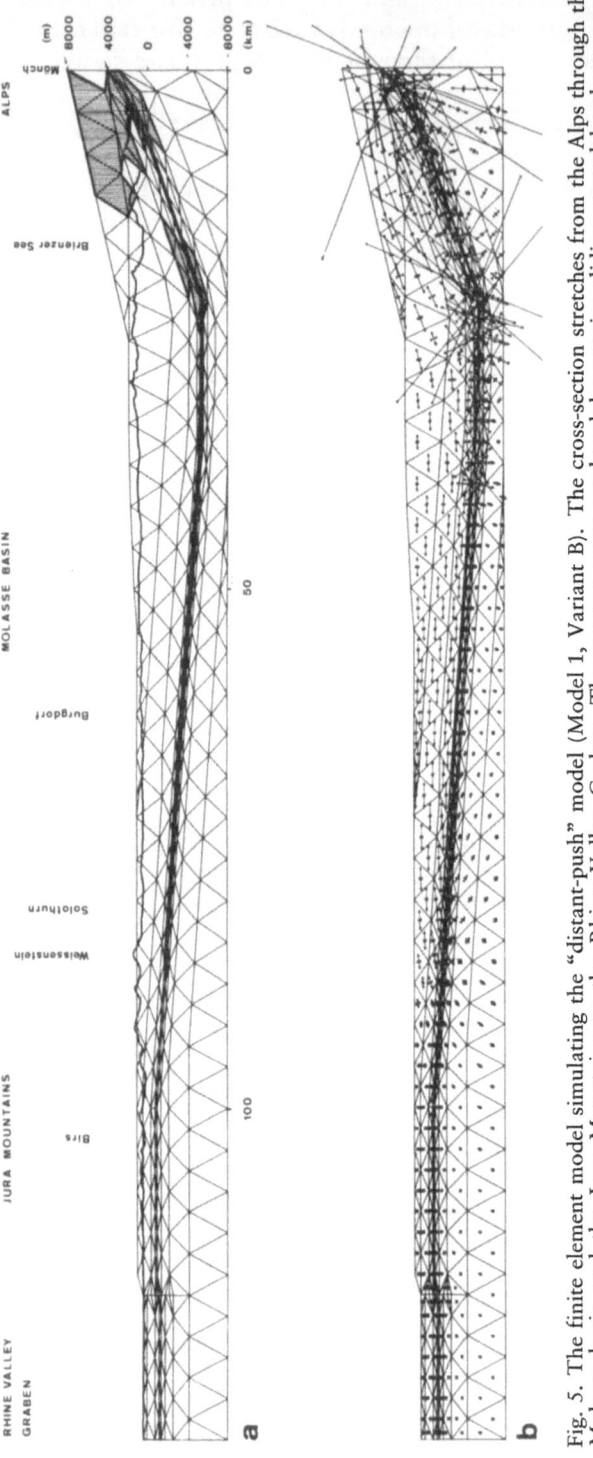

Fig. 5. The finite element model simulating the "distant-push" model (Model 1, Variant B). The cross-section stretches from the Alps through the Molasse basin and the Jura Mountains to the Rhine Valley Graben. The stresses were produced by gravity gliding caused by the unequal distribution of masses on the concave detachment horizon and by an additional push (displacement $u = 170$ m, $v = 40$ m) from the Alps. (a) The heavy line shows the development of the ductile decollement-horizon and the vertically hatched area shows the region of instability. (b) Stress distribution in the model 1, Variant B. The arrows show the direction of principal stresses and the length of arrows gives a measure of magnitude (1 mm = 3.3 kbar)

our anhydrite samples, the constants were empirically determined to be $A = 6025.6 \text{ sec}^{-1}$, $H = 27.3 \text{ kcal mol}^{-1}$, $\sigma_0 = 1700$ bar and $n = 1.5$. This flow-law is valid in the realm of a differential stress up to 3.5 kbar.

This relation states that the creep-strength of anhydrite is not a constant, but varies with strain-rate and temperature, which, in turn, is related to depth. Heat flow measurements give values corresponding to a temperature gradient of about 50^0 C per km (Von Herzen et al., 1974; Finckh, in prep.) in different Swiss lakes. However, the values determined by Rybach et al. (1978) from deep bore holes are between 30^0 C to 40^0 C per km. For our calculations we assumed a gradient of 40^0 C per km and a surface temperature of 10^0 C.

Information of the strain rate of Jura deformation is uncertain. The deformation may have taken place during the 8 million years between the late Middle Miocene and Pliocene (Bürgisser, 1979). The thickness of the anhydrite layer is about 100 m and the shortening of the overthrust slab is about 15 km. The displacement or the overthrust rate is, thus, 2 mm/y, and the average strain rate $6 \times 10^{-13} \text{ sec}^{-1}$.

Fig. 5 a illustrates the first variant of the rotational sliding model as represented by a cross-section from the Alps through the Molasse basin to the Jura mountains. The computing model assumes an elevation in the Molasse basin of about 500 m and in the Alps up to 4000 m higher than the present topography in view of the maximum possible erosion since the beginning of the Jura overthrust. This finite-element model is shared in 596 elements and 341 nodul-points. The size of the elements were reduced vertically from the base and from the surface in the direction to the detachment horizon, which should be about 100 or 200 m thick. In this model we simulated gravity sliding by a rotation of mass without an additional push from the Alps. The detachment horizon induced by boundary displacement is defined by the criterion that the stress at an element in the gliding zone is greater than the differential stress required for creep at the temperature (depth) and strain-rate assumed for that element. The results show that the detachment horizon could only be developed up to the region of Burgdorf (km 68). No other elements of the model satisfied the flow- or fracture conditions, especially not in the region of the Jura mountains, where there should be a zone of instability to induce folding by brittle rock deformation and pressure solution (Laubscher, 1976, 1979). The simulation experiment led to the conclusion that the shear force produced by gravity sliding was neither sufficient to form a decollement horizon nor enough for the building of the Jura mountains.

The second variant of the rotational sliding model is the same as the first one but with an additional push from the Alps (Fig. 5 a and b). In this model the detachment horizon could be developed up to region beneath the Jura mountains (km 88). However, Fig. 5a shows two zones of instability within the rotational slab. The first zone is in the Alps where the push is applied and the second at the deepest portion of the slab above the detachment horizon. On the other hand, a zone of instability has not been developed within the region of the Jura mountains. In other words, the

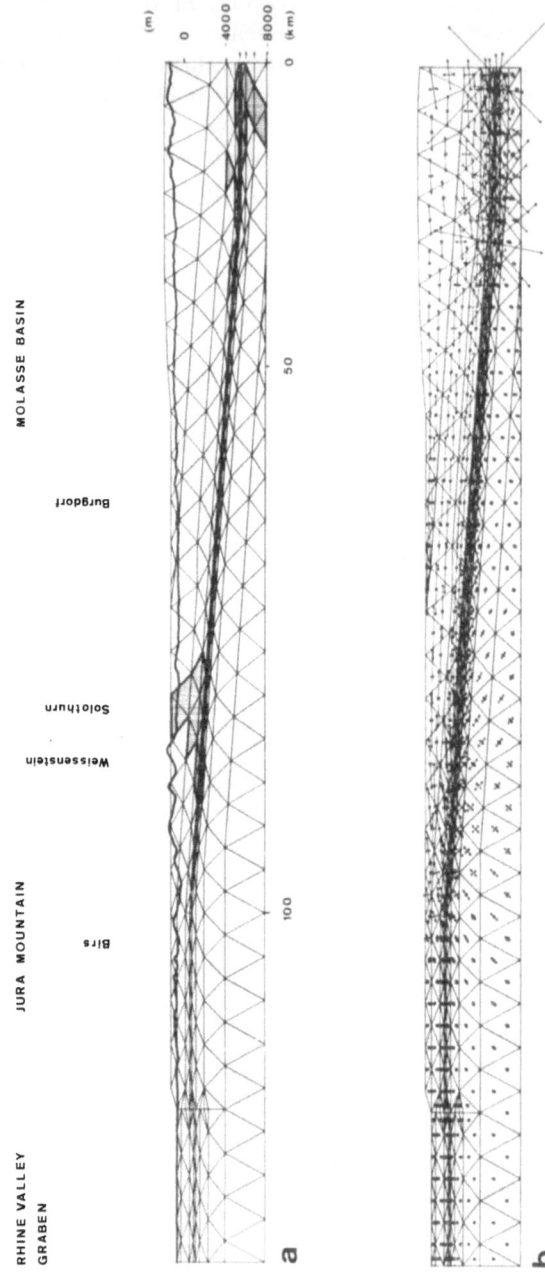

Fig. 6. The finite-element model simulating the underthrusting model (Model 2).
(a) The arrows show the direction of underthrusting (displacement $u = 225$ m, $v = 0$). The heavy line shows the development of the ductile detachment horizon and the vertically hatched area shows region of instability. (b) Stress distribution of Model 2

distant-push required to cause the stress with the gliding horizon to exceed its creep-strength is so great that the rotating slab itself would have to yield where the push was being applied, but the stress necessary to produce the Jura decollement and folding could not be transmitted by the rotating block.

In the second model we simulated an underthrusting of the Jura crystalline basement under the crystalline of the Aare Massif. The finite-element model could be shortened up to the place of underthrusting (km 22). For this simulation the whole model (Fig. 6 a) was displaced from left to right whereby the right-hand edge was fixed except for the lowest 4 nodule-points. Figs. 6 a and 6 b show the result. The anhydrite in the decollement layer was unstable up to a front well within the region of the Jura mountains (98 km); the material properties of anhydrite satisfied the conditions of instability over this long distance. Farther towards the north the detachment horizon lies too close to the surface whereby the material properties of anhydrite are similar to those of limestone (Müller and Briegel, 1978). In addition, we have two areas of instability: in the south where the crystalline basement is being underthrust, and in the region of the Jura mountains between Solothurn and Weissenstein. The overthrusting slab beneath the Molasse basin remains stable.

Another important point is that the detachment horizon develops farther to the north beyond the zone of instability within the Jura mountains. This feature of the divergence of the detachment horizon corresponds to the separation of the Folded Jura from the Table Jura. The decollement horizon under the simulated Folded Jura comes up to the surface as a thrust by brittle deformation. Meanwhile, ductile deformation under the simulated Table Jury may be responsible for the development of southern border of folds in that province. We realize the fact that the location of the separation at km 84 is south of the actually observed boundary between the two provinces. This is not surprising considering our imperfect knowledge of the strain-rate and of the temperature distribution at the time of deformation. A steeper temperature gradient or a slower strain than what we have assumed could shift the experimental boundary to coincide with what we observe in nature.

Summary and Discussion

With the application of the finite-element program "STAUB" (Kovari, 1969, 1975), it could be demonstrated that large overthrusts are unlikely within a homogeneous body. Large overthrusts need one or more weak layers in a detachment horizon. Under the Jura mountains, the weak detachment zone consists of very fine-grained anhydrite. Three different models of the Jura overthrust were calculated.

The first model assumed deformation by rotational gliding under gravity caused by the unequal distribution of masses on a concave detachment horizon. The result shows the mechanical infeasibility of this assumption: a decollement horizon cannot be developed under the Jura mountains, even if the Alps were as high as the present Himalaya mountains.

A variation of the rotational sliding model (Figs. 5 a and b) assumed an additional distant push from the Alps. Here a detachment horizon could be developed in the region under the Jura mountains. However, an area of instability was located in the Alps at the rear end of the slab where the push was being applied, but not in the region of the Jura where the slab above the detachment surface should be deformed. In other words, the stress induced by the distant push could not be transmitted by the slab to induce the Jura deformation. The best result was obtained through an assumption of the underthrusting of the Jura basement. A detachment horizon was developed to a front well within the Jura mountains, and the shear stress could be transmitted to induce instability and the deformation of the Jura, while the Molasse basin remained stable. The model also provides an explanation of the Table Jura as distinguished from the Fold Jura.

Acknowledgements

We would like to acknowledge the help given to us by the staff of the Civil Engineering Department of ETH Zürich, who allowed us the use of the STAUB program. We are particularly indebted to Dr. K. Kovari for his many helpful discussions. Mr. Bill Chapple read a first draft of this manuscript and made valuable suggestions.

References

Anderson, E. M.: The Dynamics of Faulting. Oliver and Boyd, London, 183 p. (1942).

Bock, H.: Vielfache Bruchstrukturen bei einfachen Beanspruchungen. Geol. Rundschau *61* (3), 824—849 (1972).

Bürgisser, H.: Sedimentologie der oberen Süßwasser-Molasse. Ph. D. Thesis, ETH Zürich (In prep., 1980).

Buxtorf, A.: Geologische Beschreibung des Weissenstein-Tunnels und seiner Umgebung. Beitr. geol. Karte Schweiz (NF) *21* (1907).

Clark, S. P.: Handbook of Physical Constants. Geol. Soc. Amer. Mem. *97*, 587 p. (1966).

Couples, G.: Stress and Shear Fracture (Fault) Patterns. Stress in Earth. Birkhäuser Verlag, Basel, 113—134 (1977).

Dietrich, J. H., Carter, N. L.: Stress History of Folding. Amer. J. Sci. *267*, 129—154 (1969).

Dietrich, J. H., Onat, E. T.: Slow Finite Deformation of Viscous Solids. J. Geophys. Res. *74*, 2081—2088 (1969).

Drucker, D. C., Prager, W.: Soil Mechanics and Plastic Analysis Limit Design. Quart. Appl. Math. *10*, 157—165 (1952).

Finckh, P.: Wärmeflußmessungen in Randalpenseen. Ph. D. Thesis, ETH Zurich (1976).

Finckh, P.: Heat Flow Measurements in 17 Peri-Alpine Lakes (in prep., 1980).

Fischer, H., Luterbacher, H.: Das Mesozoikum der Bohrungen Courtion 1 und Altishofen 1. Beitr. geol. Karte Schweiz (N. F.) *115*, 1—40 (1963).

Hafner, W.: Stress Distribution and Faulting. Geol. Soc. Amer. Bull. *62*, 373—398 (1951).

Handin, J.: On the Coulomb-Mohr Failure Criterion. J. Geophys. Res. *74* (22), 5343—5348 (1969).

Hsü, K. J.: Statics and Kinetics of the Glarus Overthrust. Eclog. geol. Helv. *62* (1), 143—154 (1969).

Hsü, K. J.: Thin-Skinned Plate Tectonics During Neo-Alpine Orogenesis. Am. Jour. Sci. *279*, 353—366 (1979),

Hubbert, M. K., Rubey, W. W.: Role of Fluid Pressure in Mechanics of Overthrust Folding. Geol. Soc. Amer. Bull. *70*, 115—166 (1959).

Jaeger, J. C.: Elasticity, Fracture, and Flow. Methuen, London (1969).

Kovari, K.: Ein Beitrag zum Bemessungsproblem von Untertagbauten. Schweiz. Bauzeitung *87*, 37 (1969).

Kovari, K.: The Elasto-Plastic Analysis in the Design Practice of Underground Openings. Prof. Int. Symp. on Numerical Methods in Soil Mechanics and Rock Mechanics, Karlsruhe, Sept. 1975.

Laubscher, H. P.: Die Fernschubhypothese der Jurafaltung. Eclog. geol. Helv. *54* (1), 221—282 (1961).

Laubscher, H. P.: Geometrical Adjustments During Rotation of a Jura Fold Limb. Tectonophysics *36*, 347—365 (1976).

Laubscher, H. P.: Jura Kinematics and Dynamics. Eclog. geol. Helv. *72* (2), 467—483 (1979).

Müller, W. H.: Simulation von tektonischen Überschiebungen mit Hilfe der Methode der endlichen Elemente. Ph. D. Thesis, ETH Zurich, 91 p. (1975).

Müller, W. H., Briegel, U.: Experimentelle Untersuchungen an Anhydrit aus der Schweiz. Eclog. geol. Helv. *70* (3), 685—699 (1977).

Müller, W. H., Briegel, U.: The Rheological Behavior of Polycrystalline Anhydrite. Eclog. geol. Helv. *71* (2), 397—407 (1978).

Müller, W. H., Briegel, U.: On the Mechanics of the Jura Overthrust. Eclog. geol. Helv. *73* (1) (1980).

Rybach, L., Bodmer, P., Pavoni, N., Müller, St.: Siting Criteria for Heat Extraction from Hot Rock. Pure and Appl. Geophys. *116*, 1211—1224 (1978).

Smoluchowski, M. S.: Some Remarks on the Mechanics of Overthrusting. Geol. Mag., n. s. *6*, 204—205 (1909).

Trümpy, R.: Paleotectonic Evolution of the Central and Western Alps. Bull. Geol. Soc. Amer. *71* (6), 843—908 (1960).

Trümpy, R.: Die helvetischen Decken der Ostschweiz. Eclog. geol. Helv. *62*, 105—142 (1969).

Umbgrove, J. H. F.: Origin of the Jura Mountains. Kon. Akad. Wetenschap. Nederland, Proc. *51*, 1019—1062 (1948).

Voight, B., Samuelson, B.: On the Application of Finite-Element Techniques to Problems Concerning Potential Distribution and Stress Distribution Analysis in the Earth Sciences. Pure and Appl. Geophysics *76*, 40—55 (1969).

Von Herzen, R. P., Finckh, P., Hsü, K. J.: Heatflow Measurements in Swiss Lakes. J. Geophys. *40*, 141—172 (1974).

Zienkiewicz, O. C., Cheung, Y. K.: The Finite Element Method in Structural and Continuums Mechanics. McGraw-Hill, New York (1967).

Zienkiewicz, O. C., Valliappan, S., King, J. P.: Elastoplastic Solutions of Engineering Problems' Initial Stress; Finite Element Approach. Intern. J. for Num. Method in Eng. *1*, 75—100 (1969).

Address of authors: W. H. Müller, Geological Institute, Swiss Federal Institute of Technology, ETH-Zentrum, CH-8092 Zürich, Switzerland.

Rock Mechanics, Suppl. 9, 233—244 (1980)

Rock Mechanics
Felsmechanik
Mécanique des Roches
© by Springer-Verlag 1980

Numerical Simulation of Initiating Processes of the Evolution of Sedimentary Basins: The Pannonian Basin

By

L. Bodri and **B. Bodri**

With 5 Figures

Abstract

The process of formation of sedimentary basins is considered as a result of long term vertical movements and is partitioned into two stages. The first one is essentially the formation of a depression in which water and sediments collect (the time scale of this process is of the order of 10^7 yr). The second stage is the process of isostatic adjustment (time scale of $\sim 10^8$ yr). A numerical model of the initial subsidence of sedimentary basins is presented. This model is realized by the solution of the coupled two-dimensional time dependent equations of viscous motion and of heat transport in the lithosphere-asthenosphere system characterized by non-linear rheological features. As a result of the present calculations, the subsidence of the Earth's surface or the formation of a depression can be attributed to the mechanism of a secondary convection induced mechanically by subducting lithospheric plates. Besides the formation of the depressions themselves, the present approach makes it possible to treat a number of accompanying processes, such as the uplift of hot or more dense material, the general frictional heating of the area of interest from below, e. t. c. As an illustration, the present approach is applied to the formation of the Pannonian basin.

1. Introduction

The sedimentary basins are depressions which usually do not include faulting thrusts and which are filled in with a gently dipping layer of sediments. The thickness of the sedimentary layer averages about some km; however, strong lateral variations (from zero to several km) of it can appear within the same basin. This fact necessarily implies the existence of vertical tectonic movements during sedimentation, since purely eustatic sea-level changes would not be able to produce such a variation. According to the current view, the process of formation of sedimentary basins can be partitioned roughly into two stages. As a result of intensive, mainly vertical movements, a depression develops first in which sediments collect afterwards. Subsequently, this stage is followed by isostatic adjustment under sedimentary loading.

0080-3375/80/Suppl. 9/0233/$ 02.40

While the latter stage of the evolution of sedimentary basins has been studied and is well enough understood (see e. g. Sleep and Snell, 1976, or Beaumont, 1978), the principles for the interpretation of the vertical movements taking place in the first stage have not yet been satisfactorily elaborated. Any model of evolution of sedimentary basins, however, obviously necessitates at least a hypothetical description of the forces able to induce the initial vertical movements. According to the approach that recently has become popular, the initial basin subsidence is considered in the present study as the result of a secondary convection-flow arising in areas of convergent plate margins so that it is induced in the asthenosphere by the subducting lithosphere. In this paper, a numerical model of tectonic flow and the corresponding stress and temperature fields in areas of formation of depressions is set up for different cases of plate motions and of rheological characteristics of the lithosphere-asthenosphere system. The modelling method is applied to the tectonic evolution of the Pannonian basin.

2. Lithosphere-Asthenosphere Rheology

Modelling of induced convection arising in areas of interaction of lithospheric plates requires special attention to the rheological properties of the material in the area of interest. According to Bodri and Bodri (1978, 1979), the Navier-Stokes equations of viscous motion cannot be applied for modelling the flow pattern in the whole calculation-area characterized by generally high viscosities and large viscosity variations within it. On the basis of experimental evidence, Murrell (1976) found that for acting forces of the order of 10^6—10^7 yr and mean stresses of 1—10 MPa, materials with viscosities 10^{21} Pa s or less can be considered as viscous fluids, i. e. the Navier-Stokes equations can be applied to processes taking place there. Materials with viscosities greater than or equal to 10^{23} Pa s manifest themselves as solids (elastic or in the first approximation rigid bodies). Another important point in modelling the induced convection is that some kind of non-linear (stress dependent) rheology should be introduced, otherwise one cannot obtain stable solution of the equation of motion in the highly stressed subregions near interacting plates. As probably the most reasonable approach for the description of rheological features of the viscous asthenosphere, it is assumed in the present study that the viscosity includes terms corresponding to the linear deformational process of diffusional flow and dislocation creep depending non-linearly on stress. The values of the rheological constants used for the calculation of such a viscosity are obtained for olivine by different authors and have been taken from Stocker and Ashby (1973) who reviewed this subject. Nonlinearity is taken into account also in the rigid or elastic subregions of the modelled area. These parts of the calculation — area are considered as elastic (or rigid) — perfectly plastic bodies, which means that the connection between stress and deformation is described by the Hooke's law (for a rigid body any deformation is zero) as long as the shear stress τ does not exceed some given value τ_0. Once τ_0 is reached or exceeded, the material can be transformed to plastic state.

The effects of non-linear rheology become significant in the highly stressed regions, whereas there is no remarkable difference in linear and non-linear viscosities in the case of low stresses.

3. Governing Equations and Boundary Conditions

The basic equations, numerical solution technique and boundary conditions used for modelling the induced flow, stress, and temperature fields are described in Bodri and Bodri (1978, 1979); therefore for brevity's sake, these important details are not presented here.

4. Cenozoic Evolution of the Pannonian Basin

Since the past tectonic situation in the territory of the Pannonian basin should probably have been rather complicated, the reconciliation of its tectonic history with the calculated dynamical models at the present moment may have only a schematic or preliminary character. The present work is concentrated first of all on major tectonic movements which played a dominant role in the evolution of this basin. Therefore, a number of tectonic events that occurred in the peripheral parts of the basin are left here without interpretation.

The pre-Middle-Cretaceous evolution of the Pannonian basin does not differ significantly from that of its surroundings. Its own evolution is dated from the Upper-Cretaceous period and the present state of the basin is thought to be a result of its Mio-Pliocene evolution. A number of geological and paleontological data suggest that the Pannonian basin is an area of past convergence of two microplates. The zone of collision of these plates probably should have been placed approximately in the middle of the basin with a NE−SW strike. This kind of plate convergence is much more complicated than simple subduction in the oceanic case and includes usually both, the oceanic and continental phases of subduction.

Early Stage of Oceanic Subduction

In the initial state the convergent continents are separated by an oceanic plate. As the continental blocks are getting closer, this oceanic plate, due to subduction, becomes smaller. The initial stage of the process of continental convergence, i. e. the penetration of the subducting slab into the asthenosphere, is shown schematically in Fig. 1. The penetrating slab induces large stresses on the slab-asthenosphere and slab-continental lithosphere interfaces. These stresses give rise to erosion and heating of both the slab and part of the continental lithosphere adjacent to the slab. Partial melting that occurred in the slab and in part of the asthenosphere located near the wedge-shaped region between the two plates, similarly to the case of island arc volcanism, generates andesite-volcanism. The distribution of this volcanism on the surface, i. e. the location of the chain of volcanoes, and also of the zones of probable folding and thrusting, depends almost entirely only on the form or contours of the contacting plate margins. The most intense deformations appear in those parts of the plate margins which first endure collision. Certain fragments of the convergent plate boundaries may undergo collision much later or may even not approach the collision zone at all. These parts of the plate boundaries, obviously, are characterized by much smaller stresses and deformations than the ones mentioned previously.

In the Pannonian basin the continental and oceanic plates first got into contact probably in the Upper-Cretaceous (strong folding zone about 150 km South of Budapest, with a NE—SW strike). Weak andesite-volcanism, parallel to the zone of folding but located about 100 km North of it, occurred in the Eocene. Most of the Eocene andesites are concentrated in the Zala basin, the Northern part of the Bakony mountains, the Velence mountains and also in the South-West and West of Budapest. Such a spatial

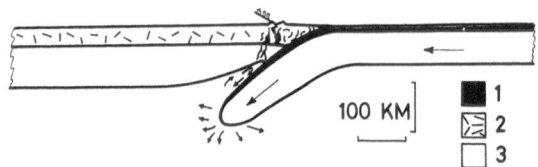

Fig. 1. Schematic sketch of the initial stage of continental convergence
(Notations: 1 — oceanic crust; 2 — continental crust; 3 — lithosphere)

distribution of these andesites possibly indicates that the slab penetrated into the asthenosphere first in the SW-part of the mentioned zone of collision of the plates. In the NE-part of the collision zone penetration occurred somewhat later as indicated by the Eocene (Lower Oligocene?) age of the Alföld flysch trough. Although the original volcanic-tectonic trends of the Paleogenic complex cannot be outlined only by its present disposition, even this disposition, however, yields some evidence for the assumption that subduction of the oceanic lithosphere occurred in the SW-part of the zone of contact of the two plates and that the dip angle of the downgoing slab had a normal value (about 40—50°).

Establishment of the Induced Convection

This stage of sedimentary basin evolution begins when the slab already has sunk to considerable depths (150—250 km). Since direct observations of the rheological features of the lithosphere, asthenosphere and the slab can give only rather uncertain results, modelling of the induced convection in the present study is carried out both for a viscous and a solid lithosphere. Due to the uncertain knowledge of the forces of interaction between the fixed continental and the subducting oceanic plates, any attempt to include an elastic lithosphere into the model makes the modelling problem rather difficult and undetermined.

Figs. 2 and 3 present the evolution of induced convection-flow with time in the case when both the asthenosphere and the lithosphere are viscous. The slab dips at 45⁰ and is given a velocity of 10 cm/yr. The high

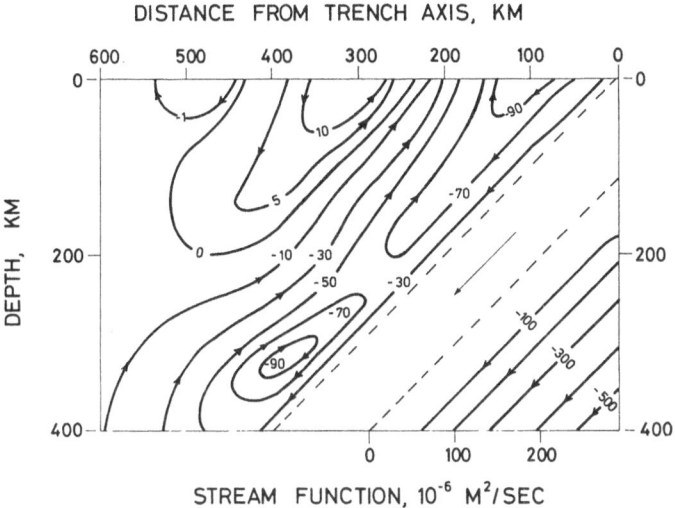

Fig. 2. Calculated flow at 1 My from the start of the calculation in the case when the whole calculation-area is viscous. The velocity of subduction V is 10 cm/yr

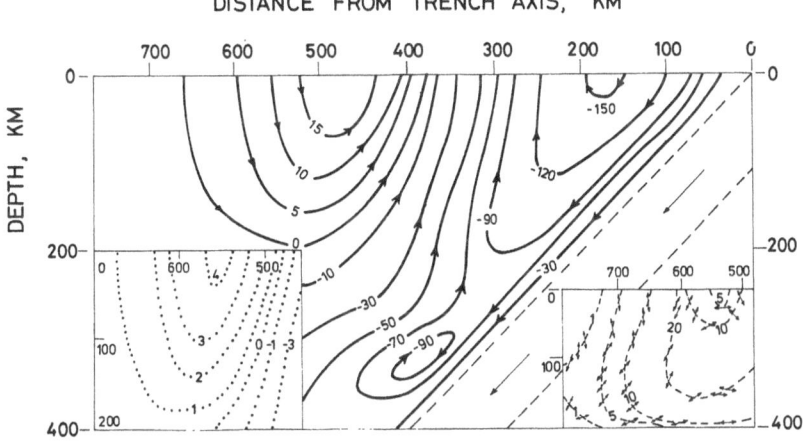

Fig. 3. Calculated stream function (solid lines, in 10^{-6} m²/sec), total shear stress (dashed lines, in MPa) and V_y flow velocity field (dotted lines, in 10^{-3} m/yr and positive sign means downward motion) at 5 My in the case when the whole calculation-area is viscous

velocity is chosen in order to maximize the effects of induced flow. Since the lithosphere and the continental crust are presented by a high-viscosity layer, the stresses and velocities can be directly computed there. For mainly numerical reasons the lithosphere viscosity is restricted to a maximum value of 10^{22} Pa s. Already at 1 My after the calculation started, induced flow in the wedge area is sufficiently established. The induced flow is relatively narrow in horizontal extent even at this high velocity of subduction. Except for a layer adjacent to the slab, the largest flow velocities ($V_y = 2$—3 cm/yr) occur along the ascending branch of the mantle flow about 300 km away from the wedge corner. At a distance about 550 km from the corner a 150 km wide zone of subsidence developes ($V_y = 0.3$—0.4 cm/yr). As time progresses, the horizontal extent of these zones remains practically unchanged. At the same time, however, a remarkable migration of their location to the periphery of the convecting area can be observed. This migration is mainly due to expansion of the near-slab branch of the convective flow. The generally insignificant temperature changes can be attributed mainly to advection. Near the slab a low-temperature zone is formed in which the material is subducted together with the slab. Hotter material is being carried towards the surface by the ascending branch of the induced flow. For 5 My the heating of this zone is no more than 200—300° K. The frictional heating in this model does not have a dominant role. It becomes more significant only in a zone of probable future volcanism of a very limited extension at the tip of the wedge-like region, whereas the slab induces sufficient flow and the stress varies from strongly compressive to strongly tensile. Since high heat flow can be observed behind most subduction zones even at considerable distances (some hundred km) from the trench, models incorporating a viscous lithosphere are not able to account for the sources of the observed heating. Excessive stresses and corresponding high frictional heating can occur only in a model with velocity boundary conditions, i. e. only if a solid lithosphere is also included in the model. This approach can give a temperature field closer to reality. On the other hand, it is less favorable from the mechanical point of view, since in this case stress and deformation in the lithosphere cannot be calculated. Large tensile stresses, however, occurring in this model at the base of the lithosphere possibly imply significant tension of the lithosphere itself and also the formation of a remarkable depression. Fig. 4 shows the induced flow pattern, the total shear stress- and temperature-fields calculated for a model incorporating a viscous asthenosphere and a rigid-plastic lithosphere at 10 My from the start of the calculation. To avoid numerical problems, the velocity of subduction is taken as 3 cm/yr. One can see in Fig. 4 that the flow pattern is much simpler here than that in the previous model. Due to the stresses exerted at the base of the solid landward plate by the lateral flow, the viscosity in a certain portion of the lithosphere decreases to such an extent that the material there can take part in the general convective motion that exists in the asthenosphere; the corresponding part of the plate is slowly subducted with the slab. As also in the previous model, a lower temperature zone occurs and expands continuously near both boundaries of the slab as time passes. Stresses of the order of 100 MPa or more are concentrated at the base of the landward plate and along the slab. These stresses are strong enough for significant frictional heating of this area to occur, and, as a result of frictional heating, a hot zone appears beneath the region where the lithosphere is narrowing. As time progresses, the temperature in this zone increases, so that by 10 My from the start of the calculation, the melting temperature

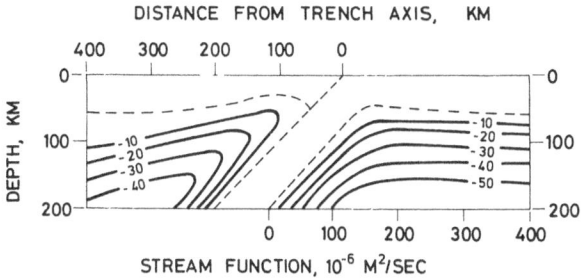

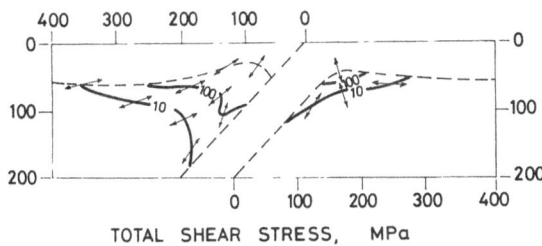

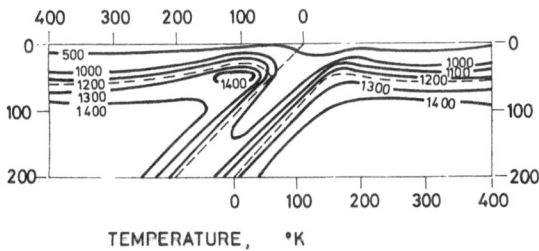

Fig. 4. Calculated flow, stress and temperature fields at 10 My for a model where the lithosphere is not viscous. V = 3 cm/yr

will have been reached in part of this zone, i. e. a source region for volcanism will have been formed. The area of plate-thinning and intense heating has a horizontal extension of about 200 km.

It should be noted that the "middle-stage" convection treated in this paragraph, which exists after the slab has sunk to middle depths, is still a small-scale phenomenon compared with the totally developed circulation which will have been established by the time the slab has already approached the high viscosity material below the asthenosphere. Therefore, the mechanical and thermal events accompanying this phase of subduction may still significantly deviate from those in island arc areas. Active back-arc spreading, formation of a marginal basin and a chain of volcanoes and also heating of the whole area even at considerable distances from the trench, even in case of high subduction velocities, occur not earlyer than at 7—10 My from the start of the subduction.

In the middle phase of subduction one can expect only more or less intense near-corner volcanism, elevation or subsidence of the lithosphere-top above ascending or the descending branches of the convection-flow and possibly breaking the lithosphere in the highly stressed regions.

As to the Pannonian basin, this middle-phase convection probably took place here in the Eocene period of its evolution. The exact separation of intra- and post-Oligocene movements in different parts of the basin is still uncertain; however, by the beginning of the Miocene the basin likely was a coherent mainland without significant variation of surface topography. The velocity of sedimentation was not too high in the Miocene and continuous sedimentation took place only in the basin-like NW-part of Hungary and also along the SW—NE trending line already mentioned above. The latter probably can be attributed to nothing but the narrow zone of subsidence near the slab as indicated in Figs. 2 and 3. A partial basin in NW-Hungary could have been formed by the other descending branch of the convective flow occurring farther from the slab (see in Figs. 2 and 3). Since the horizontal extension of the convecting region is controlled mostly by the velocity and duration of the subduction, the lithosphere-asthenosphere viscosity and also the dipping angle of the slab, the present interpretation of the location of the zones of maximum sedimentation infer a low subduction velocity and require that the dip-angle of the slab at this stage had a value of much more than 45^0. The volcanism that began already in the Paleogene, reached a maximum activity in the Middle- and Upper-Miocene, i. e. in the period of maximum development of the stage of establishment of induced convection.

Two zones of Miocene-volcanism can be distinguished spatially. The zone of ignimbrite accumulations extending from the Western part of the Mecsek mountains to NE-Hungary, including the greater part of the volcanogenic material which is strikingly elongated along the SW—NE direction, can be interpreted as the continuation of the Paleogene volcanism formed in the vicinity of the penetration of the oceanic lithosphere into the asthenosphere. The relatively small width of this zone probably also may be due to the large value of the dip-angle of the slab. Compared with the previous one, it is more problematic to interpret a small isolated zone of Middle-Miocene andesite-volcanism located in the NW-part of the basin near the inner boundary of the Western Carpathians. This zone, which is located immediately behind a basin-like region in NW-Hungary, may possibly suggest early oceanic subduction from NW that began in the Middle-Miocene. The presently observable thermal regime of the basin was probably also established during this early stage of induced convection. The mantle heat flow has maximum values also along the SW−NE direction; away from this axis is also decreases rapidly.

Tectonically, the Miocene was in general a period of block faulting which has begun in the narrow zone crossing through the central part of the Pannonian basin in a SW−NE direction already several times referred to in this paper. It is interesting to mention here the observed migration of the volcanic activity, of the block faulting and of the subsidence to the periphery of the basin. The possibility of such a migration is predicted by the present numerical calculations.

Continental Collision

During the oceanic stage of collision the major controlling factor is the quantity of oceanic lithosphere that is subducted as a downgoing slab before the continental collision. In the Pannonian basin this amount was hardly sufficient for the establishment of the last, well-developed stage of induced

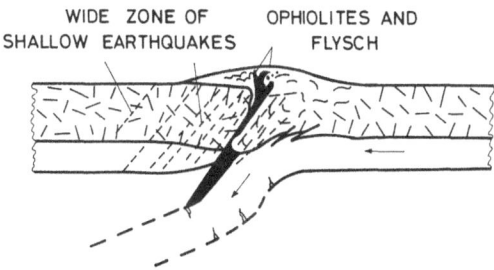

Fig. 5. Schematic sketch of continental collision

convection that exists in island arc areas. Thus, the last stage of continental collision, when the oceanic plate had already been subducted completely and the two continental blocks go into immediate contact, probably occurred in the Pannonian basin not later than in the late-Miocene or early-Pliocene periods, i. e. immediately after the transitional stage of oceanic subduction described in the previous paragraph. As is known, in a continental collision the low-density continental crust tends to prevent subduction; in case of significant compressive forces arising as a result of the resistance to subduction, this can lead to significant folding and thrusting.

Similarly to the case of oceanic subduction, the rate of junction of the continents and also the rate and time of deformations arising afterwards depend in a significant amount on the form or contours of the colliding continents. The Middle-Miocene folding in the central and Northern parts the mentioned SW−NE line probably indicates the location of the first contact of the continents. In the Southern interval of this line the continental margins had probably a weaker contact. A schematic sketch of the collision of the convergent continental margins is given in Fig. 5. At the beginning of the collision, the approaching continental margin is subducted or descends

into the trench and extrudes from there part of the oceanic crust, sediments
of the trench and also ultramafic ophiolite complexes. In the Pannonian
basin this event is indicated by interrupted ophiolite masses striking along
the whole SW−NE structural line mentioned above. This tectonic process
proceeded in a much shorter period of time than subduction of the oceanic
plate. By the end of this event a wide zone of breaking and thrusting of
the lithosphere was formed instead of one clearly developed trench where
the subduction took place. At the same time breaking and further subduction into the asthenosphere of the oceanic plate was going on. This is
probably the reason why in the whole Pannonian area inclined deep seismo-
focal zones corresponding to the Benioff zones are missing. Basin subsidence
and volcanism at this stage of evolution are practically finished. Differing
from the Miocene, the Pliocene is a period of complete ceasing of volcanism
and the final basalt volcanism began not earlier than in the Pliocene-
Quaternary boundary. The beginning of the Pliocene is also characterized
by deceleration of the basin subsidence. Apart from several minor depressions, the subsidence then was much slower or ceased in the greater part
of the Pannonian basin. The short period of collision of the continental
margins represents, essentially, the end of the active or "convecting" stage
of evolution of the basin. By the beginning of the Pliocene the basin consisted of a system of separated basin-like and elevated structures and only
at the end of the Neogene did the different subsided regions form the
uniform Pannonian basin.

Stage of the Isostatic Adjustment

The Pliocene and Quaternary in the evolution of the Pannonian basin
should probably have been periods of intense isostatic adjustment. The loading in some areas took place by sediments accumulated in certain partial
basins and by large amounts of volcanogenic material in the central sedi-
mentless part of the basin. On the basis of the work of Beaumont (1978)
on isostatic adjustment, it can be established that the development of depressions filled in with sediments of horizontal extension of about 50—100 km
proceeds in about 20 My after the initiating mechanism has ceased. Viscous
relaxation produces an increase in depth with time and a decrease in size of
the basin. The development gets continuously slower and beyond 20 My is
already very slow. The area surrounding the depression is progressively
uplifted. Since the greater part of the loading is caused by the more dense
volcanic material, the most intense subsidence had to take place in the
central part of the basin, especially in the NE-central part where most
volcanic material is concentrated. The beginning of the subsidence of the
central part of the basin is dated in the Middle-Pliocene. Approximately at
the same time some uplift of mountainous chains surrounding the Pannonian
basin occurred. As is expected, the mean velocity of sedimentation in the
Pannonian basin is somewhat increased in the Pliocene compared with that
in the Miocene and it continuously increased further in the Quaternary. If
averages are not taken over different places but sedimentation velocities

are taken at any given separate location, then the maximum velocities occur also in the Quaternary. This implies that the process of isostatic adjustment developed intensely in the Quaternary and may not yet be finished even at the present time.

5. Conclusions

The present analysis of the Cenozoic evolution of the Pannonian basin suggests that both, the initiating mechanisms and the process of isostatic adjustment are equally important factors in the formation of this basin. The present smooth topography of the basin is a result of the Pliocene-Quaternary isostatic adjustment. The initiating mechanism of formation of this basin is the collision of continental plates that took place probably in the Paleogene-Neogene.

Some important points of this collision are:

1. The collision occurred in the central part of the basin along a line with a SW − NE strike.

2. During the stage of subduction of the oceanic slab preceding the continental collision, the following events occurred:

   a) a formation of partial depressions filled in with sediments afterwards,

   b) the development of the thermal regime that exists also at present,

   c) the establishment of volcanism and igneous activity.

The subduction of the oceanic plate in the Pannonian basin proceeded along a SW − NE line and was directed towards NW.

3. The stage of oceanic subduction is much shorter here than that in the case of the island arcs. This is due to the smaller amount of material of the oceanic lithosphere plate. Therefore all dynamic processes arising here during this stage appear relatively close to the convergence zone.

4. The process of continental collision is much shorter than the oceanic subduction and it is indicated by the formation of a suture zone where continental crust approaches the trench and also by deformation and breaking of the subducted part of the oceanic lithosphere.

References

Beaumont, C.: The Evolution of Sedimentary Basins on a Viscoelastic Lithosphere: Theory and Examples. Geophys. J. R. Astron. Soc. 55, 471—497 (1978).

Bodri, L., Bodri, B.: Numerical Investigation of Tectonic Flow in Island-Arc Areas. In: M. N. Toksöz (ed.), Numerical Modeling in Geodynamics. Tectonophysics 50, 163—175 (1978).

Bodri, L., Bodri, B.: Flow, Stress and Temperature in Island Arc Areas. Geophys. Astrophys. Fluid Dynamics 13, 95—106 (1979).

Murrell, S. A. F.: Rheology of the Lithosphere — Experimental Indications. Tectonophysics 36, 5—24 (1976).

Sleep, N. H., Snell, N. S.: Thermal Contraction and Flexure of Mid-Continent and Atlantic Marginal Basins. Geophys. J. R. Astron. Soc. *45*, 125—154 (1976).

Stocker, R. L., Ashby, M. F.: On the Rheology of the Upper Mantle. Revs. Geophys. Space. Phys. *11*, 391—426 (1973).

Address of authors: Dr. L. Bodri, B. Bodri, Department of Geophysics, Eötvös University, Kun Bela ter 2, H-1083 Budapest, Hungary.

Rock Mechanics, Suppl. 9, 245—255 (1980)

Rock Mechanics
Felsmechanik
Mécanique des Roches
© by Springer-Verlag 1980

Reconstruction of Stress Fields for the Aegean by a Finite Element Model

By

M. Shulman and W. Skala

With 6 Figures

Abstract

Since Ritsema 1974, a large number of focal plane data exist for the Aegean. In the present study, the reconstruction of the stress field due to the direction of maximum compression as found in Ritsema is attempted by a Finite-Element model. The fracture patterns thus simulated are compared with the results of photolineations determined from aerial photos and satellite images as well as bathymetric and reflection — seismic investigations, in order to determine whether a relation exists between the recent or subrecent fracture pattern and the recent stress distribution.

1. Introduction

The current geodynamic situation of the Hellenides has been the subject of numerous geophysical and geological investigations in the last few years. Thus, detailed investigations into the distribution of gravity anomalies and the structure of the crust have been carried out (Makris, 1978). The relative high seismicity of the southern Aegean has been explained by Caputo et al., 1970. Papazachos, 1973. Morelli et al., 1975. Richter and Strobach, 1978. Leydecker et al., 1978, among others as due to underthrusting of the European plate by the African plate along a Benioff zone dipping northward from Crete which has furthermore been related to the South Aegean volcanic arc (Makris, 1976; Jacobshagen et al., 1978). Makris (1978) deduced from the results of geothermal studies in the Aegean (Jongsma, 1974), a thermal regime that corresponds approximately to a geothermal model of an oceanization according to Schuiling (1969). On the basis of focal plane analyses conducted by Constantinescu et al. (1966), Ritsema (1970, 1977), and Strobach and Richter (1978), Jacobshagen et al. (1978) deduced the following: compression striking NE along the outer part of the Hellenic arc and tension diverging in the central region.

Makris (1976, 1978) developed a lithothermal model based on the activity of a hot plume in the central Aegean. This was induced by the relative movement of Africa and Europe towards one another, whereby

0080-3375/80/Suppl. 9/0245/$ 02.20

crustal shortening and subduction appeared. This model forms essentially the basis of the model put forth by Jacobshagen et al. (1978) who assume that a new orogenic belt is just developing outside the Hellenic arc. The present geodynamic processes in the interior of this arc characterize the evolution of the young back-arc basin. From this point of view, the young updoming of the South Aegean area is interpreted as induced by foregoing subduction. Thermodiapirism is not believed to be the source of recent orogenesis.

A deviating model of the recent geodynamic development of the Aegean was proposed by Brunn (1976): The convexity of the Hellenic arc towards the foreland has been induced, according to Brunn, by a relative westward movement of the Arabian plate. Jacobshagen et al. (1978) also deduced [from the focal plane data of Ritsema (1974)] compression running E − W in the North Aegean that can be related to westward movement of the Anatolian block along the North-Anatolian lineament relative to Europe. They build with this model upon a suspicion put forth by Papazachos (1973, 1976).

With regard to the direction of movement, the ideas laid down by Brunn (1976) partially correspond to the plate-tectonic reflections of McKenzie (1972, 1977) and McKenzie et al. (1970), who believed that they recognized in the Aegean a SW-movement of the Aegean plate and a westward movement of the Turkish plate; the westward movement of the Turkish plate being caused by a northward movement of the Arabian plate. Papazachos and Delibasis (1969) also inferred from the analysis of focal plane data phenomena that currently strengthen the convexity of the Hellenic arc. From the distribution of focal plane solutions, they gained the impression of a bent and extended beam surrounding the relatively stable mass of the Central Aegean Sea. In our opinion, however, the results as presented by Papazachos and Delibasis (1969) do not allow such an interpretation since the stress distribution predicted by beam theory cannot be ascertained from the focal plane analysis. McKenzie (1972, 1977) assumes essentially a rigid plate and considers the two dimensions parallel to the Earth's surface. The vertical component of the deformation in the central Aegean induced by a westward movement of Anatolia, has been concluded by Brunn (1977) and by Jacobshagen et al. (1978).

According to Brunn (1977), the convexity of the Hellenic arc was already induced in the Eocene. Jacobshagen et al. (1978), on the other hand, are of the opinion that the present geodynamic situation in the Aegean, based on geologic data, can only be traced back as far as the Pliocene. Makris (1976) correlates temporally the origin of the Benioff zone with the initiation of the South Aegean volcanic arc.

2. Fracture Pattern and Stress Field in the Aegean

The attempts just mentioned to reconstruct the general direction of movement of the Earth's crust in the Aegean are based primarily on interpretations of focal plane data [Ritsema (1974), McKenzie (1972, 1974), etc.]. A precise knowledge of the fracture pattern is therefore of great significance in ascertaining the stress field. For the Aegean, results have been presented by Bingol (1976), Foose (1977), Kronberg and Gunther (1977), Jacobshagen et al. (1978), Letouzey et al. (1977), Angelier (1978), Angelier and Le Pichon (1978), and Meissner (1979). While Bingol (1976) summarized all previous results of conventional field investigations, Foose (1977), Kronberg and Gunther (1977), and Letouzey et al. (1977) limited themselves to the evaluation of satellite photos (Landsat-images). Angelier (1978) and Angelier and Le Pichon (1978) utilized basically geologic field measurements, complemented, however, by focal plane solutions and bathymetric data. Currently, the most comprehensive survey has been compiled by Meissner (1979) whose fracture pattern encompasses the following: He combined field investigations with large-scale aerial photo interpretations and the evaluation of satellite images and included the results of oceanic seismic investigations (Jongsma, 1975, and Jongsma et al., 1977). Meissner et al. (1979) made a preliminary compilation of these results and found that at most two sets of fractures had been formed locally in the Aegean. On the basis of this observation a model of conjugate strike slip faults with σ_1 and σ_3 parallel to the earth's surface could be developed, which could be confirmed by geologic and geophysical field data. The bisectrix of the acute angle formed by the conjugate fractures correspond to the local direction of the maximum compressive stress. The stress pattern derived from the fracture pattern was congruent with the convex arc of the southern Hellenides.

Angelier and Le Pichon (1978) determined, on the other hand, that in the southern part of the Aegean, above all in the region of Crete, normal faults predominate, while, in the central and northern Aegean strike-slip faults predominate. They deduced from it a N−S tension that had been caused by a relative westward movement of Anatolia. If one compares the conclusions of Angelier and Le Pichon (1978) with those of Jacobshagen et al. (1978) or of Brunn (1977), the following becomes evident: the westward movement of Anatolia is an integral part of all the hypotheses seeking to explain the current geodynamic situation in the Aegean and is supposed to have influenced the convexity of the arc. While Angelier and Le Pichon (1978) claim to recognize traces of tension in the earth's crust in the outer part of the arc and compression in its center, Brunn (1977) observed tensile phenomena within the arc including recent normal faults in the Aegean Sea, and compressive phenomena along the border. However, both come to comparable conclusions despite their different interpretations. As the recent investigation of Meissner (1979) shows, both tensile and compressive phenomena can be observed within narrow confines in many areas of the middle Aegean. He could, for example, localize recent normal faults cut by strikeslip faults. These occurrences point to the necessity of considering the recent tectonics of the Aegean as polyphasal. This has been done by Angelier (1978). Within the time span considered here, he differentiates a Plio-Pleistocene and a Pleistocene-Holocene tensile phase as well as a Pleistocene compressive phase. According to Jacobshagen et al. (1978) the recent structural evolution of the Aegean has been determined by a Pleistocene tensile phase and a Pleistocene compressive phase.

3. Internal Stress Distribution and External Loading of the Aegean

As shown, Brunn (1977), Angelier (1978), Jacobshagen et al. (1978), and Angelier and Le Pichon (1978) have attempted to deduce possible directions of external loading of the Aegean from fracture patterns or assumed stress distributions. Makris (1978) and Strobach and Richter (1978) have also made corresponding attempts at reconstruction based on the results of their geophysical investigations. It must, however, be pointed out that a unique reconstruction of the external loading of bodies whose stress distributions are known, is not possible. It is therefore not surprising that the numerous attempts to reconstruct the geodynamic history of the Aegean have led to strongly diverging hypotheses. Meissner et al. (1979) showed, for example, different stress patterns corresponding to different constitutive equations in a finite element model of the Aegean, although the external loading remained the same. Nevertheless, general estimates of the external loading must be taken seriously as an indication of possible large-scale relative movements, if other geologic indices are included. Such estimates have been made by Angelier (1978), Brunn (1977), Jacobshagen et al. (1978), etc. If the hypothesis requires three dimensions, such as Brunn (1977), then it is no longer possible to verify it exactly. A polyphasal tectonics as proposed by Jacobshagen et al. (1978) and Angelier (1978) could be explained by temporally changing external loading.

Other than small-scale laboratory experiments, mathematical models offer us the opportunity to examine these hypotheses more closely. Due to its ease in handling boundary conditions, the finite element method has become more and more the standard numerical method for solving differential equations and, in particular, stress-strain problems. This method was utilized in their mathematical model of the Aegean by Meissner et al. (1979). In the majority of their simulations, displacements were assigned along the boundary and corresponded to the P-axes tabulated by Ritsema (1974). The direction of the maximum compression ran generally parallel to the strike of the arc. Thus, agreement between the simulated stress pattern and the young fracture pattern appeared to exist. This investigation was, however, carried out to determine whether the recent fracture pattern in the Aegean could be correlated with underthrusting of the African plate.

As previously mentioned, the geodynamic hypotheses found in the literature are based on a suspected westward movement of Anatolia. This westward movement should exert an influence on the fracture pattern in the Aegean. This fact forms the basis of the following finite element investigation.

4. Finite Element Calculations

A semicircle with a radius of 624 units was chosen as a first approximation of the Hellenic arc. The model consists of 497 nodal points and

150 isoparametric elements. An elastic constitutive law was used. Since no free surface is allowed along the perimeter of the arc, the following method was applied to simulate the boundary conditions (see Fig. 1): along AC the nodal points are allowed to move in the Y direction only. The nodal points along AOB were held constant, while the westward movement of the Turkish plate was simulated by BC moving in the $-Y$ direction. The nodal forces X_f, Y_f necessary to insure no displacement along AOB were then calculated and used as input data for further calculations. The nodal forces were modified in three ways: (1) The force $F = (X_f{}^2 + Y_f{}^2)^{1/2}$ was multiplied by a factor R, while the angle $\Theta = \tan (Y_f/X_f)$ was held constant. (2) F was held constant and Θ varied. (3) Both F and Θ were varied. Use of no forces eliminates the free surface and enables a resistance to be specified along any segment of AOB.

Since only two-dimensional calculations were carried out, a fracture pattern based on the resulting stress fields requires an estimate of the vertical stress. If plane strain conditions are assumed, the second principal stress is given by the expression

$$\sigma_2 = \nu\,(\sigma_1 + \sigma_3)$$

where ν is Poisson's ratio and $\sigma_1 > \sigma_2 > \sigma_3$. The vertical stress must then be equal to the intermediate stress, thus allowing only strike-slip faults to occur. If the vertical stress is assumed to be σ_1, normal faults result. In this case the strike of the faults corresponds to the direction of the intermediate stress. Reverse faults can be expected if the vertical stress is σ_3. The complexity of the fracture pattern in the Aegean points to a strong dependence on the magnitude of the vertical stress.

Fig. 1 shows the stress field for the boundary condition allowing no displacement along AOB. The minimum principal stresses strike generally NW−SE resulting in either strike-slip faults approximately E−W or N−S or NE−SW striking normal faults.

In Fig. 2 the nodal forces were set zero except for three nodal points at the left hand side where no displacement was allowed. The minimum principal stresses strike N−S. Thus normal faults can be expected to strike E−W, while strike-slip faults strike either NW−SE or NE−SW.

Fig. 3 shows the stress field when the nodal forces along the boundary are altered to yield an angle of $\Theta = 215$. The variation of Θ caused the largest change in the nodal forces to occur in the southwest quadrant. Two areas can be distinguished: a central area striking NW−SE and an outer area exhibiting a radial pattern varying from a direction of NE−SW in the southwest quadrant to NW−SE in the southeast quadrant. The fracture pattern now shows a break in the strike direction in the southwest quadrant where the central and outer areas meet.

Maintaining the same magnitude of nodal forces as in Fig. 3, Fig. 4 shows the stress field resulting from rotating Θ between O and B to 135. The symmetry of the boundary conditions is reflected by the stress field.

In Fig. 5 the magnitude of the nodal forces along AOB was doubled and the sign of the Y_f nodal force between O and B changed. The rotation of Θ to 135 in Fig. 4 also caused the sign of Y_f to change, and, in addition, the magnitudes of X_f and Y_f to change; Figs. 4 and 5 differ however in Θ and F. The increase in the magnitude of the nodal forces influences the extent of the penetration of the outer area.

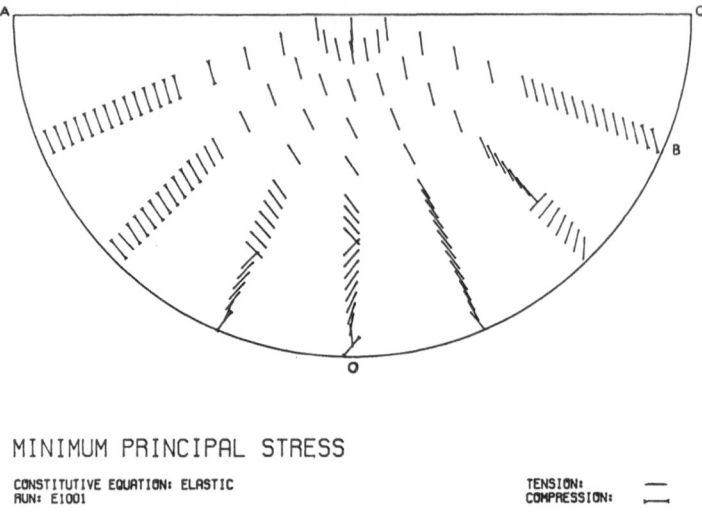

MINIMUM PRINCIPAL STRESS

CONSTITUTIVE EQUATION: ELASTIC   TENSION: —
RUN: E1001   COMPRESSION: ⊢—

Fig. 1. Stress field due to no displacement along AOB

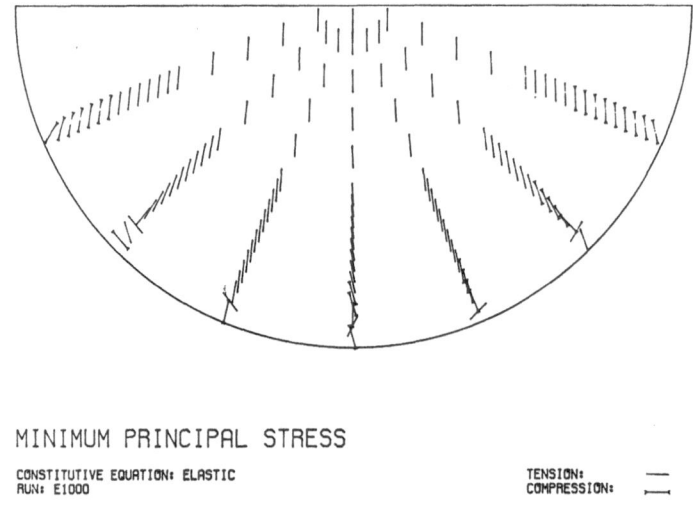

MINIMUM PRINCIPAL STRESS

CONSTITUTIVE EQUATION: ELASTIC   TENSION: —
RUN: E1000   COMPRESSION: ⊢—

Fig. 2. Stress field due to free surface

Fig. 6 was the only simulation where displacements instead of nodal forces were assigned along AOB. Thus a shortening of the semicircle took

place. In Figs. 2—5, on the other hand, an elongation of the semicircle occurred. The displacements were always perpendicular to the perimeter of

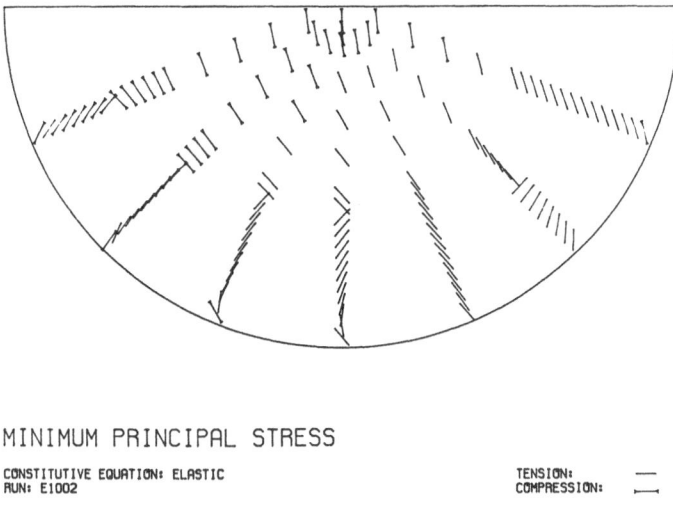

MINIMUM PRINCIPAL STRESS

CONSTITUTIVE EQUATION: ELASTIC
RUN: E1002

TENSION: ——
COMPRESSION: ⊢—⊣

Fig. 3. Stress field due to Θ being set to 215⁰ along AOB

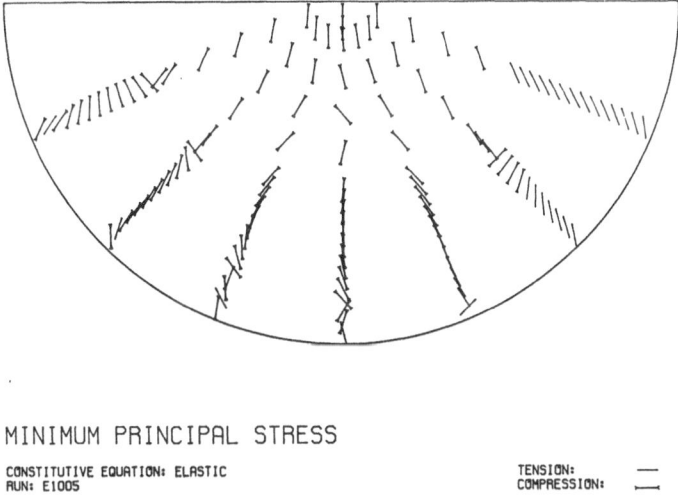

MINIMUM PRINCIPAL STRESS

CONSTITUTIVE EQUATION: ELASTIC
RUN: E1005

TENSION: ——
COMPRESSION: ⊢—⊣

Fig. 4. Stress field due to Θ being set to 215⁰ along AO and 135⁰ along OB

the semicircle and were chosen to be half the value of the displacements
along BC. This symmetry of the boundary conditions is reflected in the
stress pattern.

5. Discussion

Although the westward movement of Anatolia was simulated in all calculations by BC moving in the −Y direction, the boundary conditions along AOB influence considerably the stress pattern. The first two examples (Fig. 1 and 2) represent extreme cases that cannot be expected to be found in nature. Of special interest, however, is Fig. 3 that can be looked at as

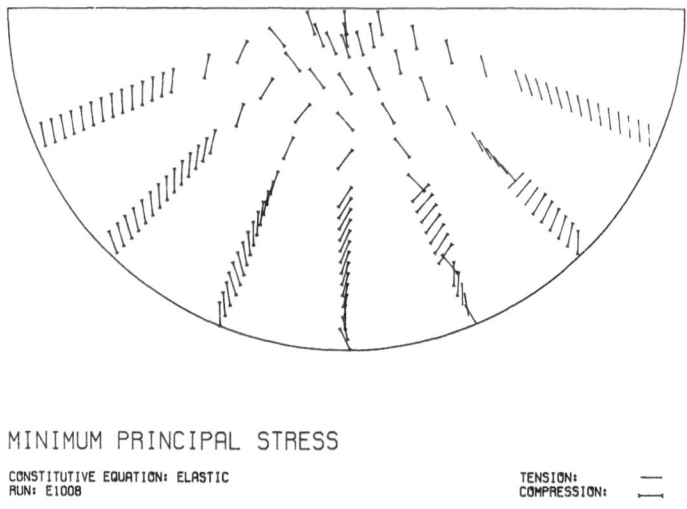

MINIMUM PRINCIPAL STRESS

CONSTITUTIVE EQUATION: ELASTIC   TENSION: ———
RUN: E1008   COMPRESSION: ⊢———⊣

Fig. 5. Stress field due to nodal forces being doubled along AOB and a change of Θ along OB

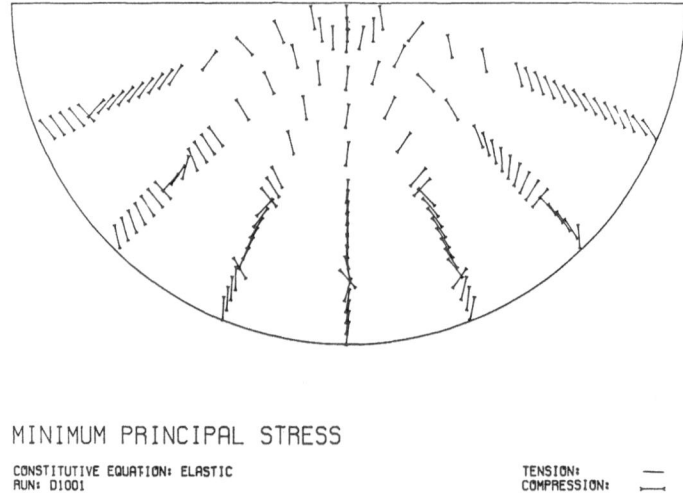

MINIMUM PRINCIPAL STRESS

CONSTITUTIVE EQUATION: ELASTIC   TENSION: ———
RUN: D1001   COMPRESSION: ⊢———⊣

Fig. 6. Stress field due to displacements along AOB being set to half the displacement along BC

an interaction between a westward movement of Anatolia and a northeastward movement caused by a Benioff zone dipping to the northeast in the southern Aegean. The maximum principal stresses exhibit a radial pattern such as demanded by Meissner et al. (1979). From the pattern of minimum principal stresses, strike-slip faults in the eastern part of the Aegean and normal faults in the northern and western parts can possibly be deduced. Corresponding relations can be recognized in Figs. 4 and 6, albeit with different dimensions. These patterns call to mind observations applied by Brunn (1977) as the basis of his reconstruction. It appears, however, difficult to reconcile the results of the simulations with the ideas of Angelier and Le Pichon (1978). The polyphase tectonics in the Aegean such as assumed by Angelier (1978) and Jacobshagen et al. (1978) can be explained as the overlapping of individual tectonic events, for example Figs. 3, 4 and 6. The complicated recent fracture pattern of the Hellenides can be traced back to a westward movement of Anatolia, as long as the Hellenides are subject to temporally changing boundary conditions.

6. Summary

Finite element calculations show that the recent fracture pattern in the Aegean can be caused by a westward movement of Anatolia. Several movement phases separated temporally and subject to different boundary conditions would however be necessary.

References

Angelier, J.: Tectonic Evolution of the Hellenic Arc Since the Late Miocene. Tectonophysics 49, 23—36 (1978).

Angelier, J., Le Pichon, X.: The Hellenic Arc, a Key to the Kinematics of the Eastern Mediterranean Since 13 Million Years. XXVIe Congrès-Assemblée plénière CIESM, 2 p., 1 fig., Antalya (1978).

Bingöl, E.: Evolution géotectonique de l'Anatolie de l'Ouest. Bull. Soc. géol. France 18, 431—450, Paris (1976).

Brunn, J. H.: L'arc concave Zagro-taurique et les arcs convexes taurique et égéen: collision et arcs induits. Bull. Soc. géol. France (7), 18, 553—567, 5 fig., Paris (1976).

Brunn, J. H.: Über die Entstehung gefalteter Ketten: Kollisionstektonik und induzierte Bögen. Z. dt. geol. Ges. 127, 323—335 (1976).

Caputo, M., Panza, G. F., Postpischl, D.: Deep Structure of the Mediterranean Basin. J. Geophys. Res. 75, 4919—4923 (1970).

Constantinescu, L., Ruptrechtova, L., Enescu, D.: Mediterranean-Alpine Earthquake Mechanisms and Their Seismotectonic Implications. Geophys. J. Roy. Astron. Soc. 10, 347 (1966).

Foose, R. M.: Structural Lineaments and Tectonics of the Mediterranean Basin. Intern. Sympos. Split CIESM, 221—232, Ed. Technip, Paris (1977).

Jacobshagen, V., Dürr, St., Kockel, F., Kopp, K.-O., Kowalczyk, G.: Structure and Geodynamic Evolution of the Aegean Region. Alps, Apennines, Hellenides, Inter-Union Comm. Geodynm., Sci. Rept. *38, 537—564*, 8 fig., Schweizerbart, Stuttgart (1978).

Jongsma, D.: A Marine Geophysical Study of the Hellenic Arc. Dissertation, 69 pp., 42 fig., Cambridge (1975).

Jongsma, D., Wissman, G., Hinz, K., Garde, S.: Seismic Studies in the Cretan Sea. The Southern Aegean Sea: An Extensional Marginal Basin Without Sea-Floor Spreading? Meteor Forsch. Ergeb., Reihe C *27, 3—30*, 13 fig., Borntraeger, Berlin – Stuttgart (1977).

Kronberg, P., Günther, R.: Fracture Patterns and Principles of Crustal Fracturing in the Aegean Region. VI. Coll. Geology Aegean Region, Proceedings *2, 893—906*, 12 fig., Athen (1977).

Letouzey, J., Tremolieres, P., Biju-Duval, B.: An Approach to the Structure of the Mediterranean Area: A Satellite Photogeological Study. Intern. Sympos. Split CIESM, *215—220*, 1 fig., Ed. Technip, Paris (1977).

Leydecker, G., Berckhemer, H., Delibasis, N.: A Study of Seismicity in the Peloponnesus Region by Precise Hypocenter Determinations. Alps, Apennines, Hellenides, Inter-Union Comm. Geodynm., Sci. Rept. *338, 406—410*, 2 fig., Schweizerbart, Stuttgart (1978).

Makris, J.: A Geodynamic Model of the Hellenic Arc Deduced from Geophysical Data. Tectonophysics *36, 339—346* (1976).

Makris, J.: A Geophysical Study of Greece Based on: Deep Seismic Soundings, Gravity, and Magnetics. Alps, Apennines, Hellenides, Inter-Union Comm. Geodynm., Sci. Rept. *38, 392—401*, 7 fig., Schweizerbart, Stuttgart (1978).

McKenzie, D. P.: Active Tectonics of the Mediterranean Region. Geophys. J. Roy. Astron. Soc. *30, 109—181* (1972).

McKenzie, D. P.: Can Plate Tectonics Describe Continental Deformation? Intern. Sympos. Split CIESM, *189—196*, 7 fig., Ed. Technip, Paris 1977.

Meissner, B.: Bruchtektonik in der Zentral-Ägäis. Berliner Geowiss. Abhandlungen (in press) (1979).

Meissner, B., Shulman, M., Skala, W.: Finite-Element-Berechnungen zur Auswertung herdmechanischer Daten und deren Beziehungen zum Störungsmuster des Ägäis-Raumes. Geol. Rdsch. *68, 225—235* (1979).

Morelli, C., Pisani, M., Gantar, C.: Geophysical Studies in the Aegean Sea and the Eastern Mediterranean. Boll. di Geofisica teor. ed appl. *18, 127—168* (1975).

Papazachos, B. C.: Distribution of Seismic Foci in the Mediterranean and Surrounding Area and its Tectonic Implication. Geophys. J. Roy. Astron. Soc. *33, 421—430* (1973).

Papazachos, B. C., Delibasis, N. D.: Tectonic Stress Field and Seismic Faulting in the Area of Greece. Tectonophysics *7, 231—255* (1969).

Richter, I., Strobach, K.: Benioff Zones of the Aegean Arc. Alps, Apennines, Hellenides, Inter-Union Comm. Geodynm., Sci. Rept. *38, 410—414*, 6 fig., Schweizerbart, Stuttgart (1978).

Ritsema, A. R.: Seismo-Tectonic Implications of a Review of European Earthquake Mechanisms. Geol. Rdsch. *59, 36—56* (1970).

Ritsema, A. R.: The Earthquake Mechanisms of the Balkan Region. UNESCO Survey Seismicity Balkan Region, UNDP Project REM/70/172, 36 pp., 26 fig., De Bilt (1974).

Schuiling, R. D.: A Geothermal Model of Oceanization. Verhandl. Kon. Ned. Geol. Mijnbouwk. Gen. *26*, (1969).

Address of authors: M. Shulman, Institut für Kerntechnik, Technische Universität Berlin, Marchstrasse 18, D-1000 Berlin 12; W. Skala, Institut für Geologie, Freie Universität Berlin, Altensteinstrasse 34a, D-1000 Berlin 33.